MARINE GEOLOGY

MARINE GEOLOGY

UNDERSEA LANDFORMS AND LIFE FORMS

The Changing Earth Series

JON ERICKSON

Facts On File, Inc.
AN INFOBASE HOLDINGS COMPANY

MARINE GEOLOGY: UNDERSEA LANDFORMS AND LIFE FORMS

Facts On File, Inc.
11 Penn Plaza
New York, NY 10001

Library of Congress Cataloging-in-Publication Data

Erickson, Jon, 1948–
Marine geology: undersea landforms and life forms / Jon Erickson.
p. cm. — (The changing earth series)
Includes bibliographical references and index.
ISBN 0-8160-3354-4
1. Submarine geology. 2. Marine biology. I. Title. II. Series: Erickson, Jon, 1948– Changing earth.
QE39.E68 1996
551.46'08—dc20 96-22109

Text design by Ron Monteleone/Robert Yaffe
Jacket design by Catherine Rincon Hyman
Printed in the United States of America

RRD FOF 10 9 8 7 6 5 4 3 2 1

This book is printed on acid-free paper.

CONTENTS

TABLES

ACKNOWLEDGMENTS

The author thanks the following organizations for providing photographs for this book: the National Aeronautics and Space Administration (NASA), the National Oceanic and Atmospheric Administration (NOAA), the U.S. Army Corps of Engineers, the USDA Forest Service, the USDA Soil Conservation Service, the U.S. Defense Nuclear Agency, the U.S. Department of Energy, the U.S. Geological Survey (USGS), the U.S. Maritime Administration, the U.S. Navy, and the Woods Hole Oceanographic Institution (WHOI).

INTRODUCTION

Our planet contains so much water that perhaps it would have been better named Oceania. Ours is the only known body in the Solar System that is surrounded by water filled with unique geologic structures and teeming with a staggering assortment of marine life forms. Some of the strangest creatures on Earth, whose ancestors go back several hundred million years, live on the deep ocean floor. Much of the world's untapped wealth lies under the waves. And the seabed is a new frontier for the discovery of mineral resources.

The floor of the ocean features a rugged landscape unmatched anywhere on the continents. Vast undersea mountain ranges much more extensive than those on land crisscross the seabed. Although deeply submerged, the midocean ridges are easily the most pronounced features on the planet, extending over an area larger than that covered by all the major terrestrial mountain ranges combined. The ocean floor is continuously being created at spreading ridges, where molten rock oozes out of the mantle, and being destroyed in deep-sea trenches off the continents and island arcs in the open ocean. The subduction of the ocean crust in deep-sea trenches plays a fundamental role in global tectonics and accounts for powerful geologic forces that continuously shape the planet.

An extraordinary number of volcanoes are hidden under the waves, and most of the volcanic activity that continually remakes the surface of the Earth occurs on the ocean floor. Active volcanoes rising up from the bottom of the ocean create the tallest mountains, dwarfing even those on the continents. Most of the world's islands in fact began as undersea volcanoes that broke the surface of the sea. However, the preponderance of marine volcanoes are not exposed at the surface and remain as isolated seamounts. Many ridges host an eerie world that time forgot, a cold, dark abyss comprised of tall chimneys spewing hot, mineral-rich water that support unusual animals previously unknown to science.

Rivers of strong, flowing ocean currents are the main transport system for distributing water and heat to all corners of the world. Chasms that challenge the largest terrestrial canyons plunge to great depths. Massive submarine slides gouge deep depressions into the seabed and deposit enormous heaps of sediment on the ocean floor. Undersea slides also occasionally generate tall waves that pound nearby shores, causing much destruction to seaside communities. Abyssal storms with strong currents sculpt the ocean bottom, churning up huge clouds of sediment, dramatically modifying the seafloor. The scouring of the seabed and the deposition of large amounts of sediment results in a highly complex marine geology.

The most intriguing terrain features exist on the bottom of the ocean. The ocean floor hosts a myriad of unique geologic formations. Unusual seamounts erupt mud instead of lava like most volcanoes. Scattered along the seafloor are remarkable volcanic deposits, including piles of pillow lavas, forests of black and white smokers, and undersea geysers, whose warm, mineral-laden water rises to the surface in massive plumes. The active undersea world sports a variety of sea caves, blue holes, calderas, and craters formed by undersea explosions and meteorite impacts. These are just a few examples among the many wonders of the seabed.

1

THE BLUE PLANET

Throughout our planet's long history, as many as 20 oceans have come and gone, as continents drifted apart and reconverged into supercontinents. The present ocean basins formed after a supercontinent named Pangaea, Greek meaning "all lands," broke apart into today's continents about 170 million years ago. Before the breakup, a single large ocean called Panthalassa, Greek meaning "universal sea," surrounded the supercontinent. Prior to the assembly of Pangaea, all continents surrounded an ancient Atlantic Ocean called the Iapetus Sea. Deeper into the past, the continents again formed a supercontinent, and its breakup created entirely new seas, which participated in a great explosion of life. Life itself possibly evolved at the bottom of a global ocean not long after the Earth was created.

ORIGIN OF SEA AND SKY

During the Earth's formative years, a barrage of asteroids and comets pounded the infant planet and its moon (Fig. 1–1). Some meteorites were stony, with rock and metal; others were icy, with frozen gases and water ice; many contained carbon, as though coal rained down from the heavens. It is possible that these carbon-rich meteorites bore the organic molecules

TABLE 1–1 THE GEOLOGIC TIME SCALE

Era	Period	Epoch	Age (millions of years)	First Life Forms
		Holocene	0.01	
	Quaternary			
		Pleistocene	2	Man
Cenozoic		Pliocene	7	Mastodons
		Miocene	26	Saber-tooth tigers
	Tertiary	Oligocene	37	
		Eocene	54	Whales
		Paleocene	65	Horses Alligators
	Cretaceous		135	
				Birds
Mesozoic	Jurassic		190	Mammals
				Dinosaurs
	Triassic		250	
	Permian		280	Reptiles
	Pennsylvanian		310	
				Trees
	Caboniferous			
Paleozoic	Mississippian		345	Amphibians Insects
	Devonian		400	Sharks
	Silurian		435	Land plants
	Ordovician		500	Fish
	Cambrian		570	Sea plants Shelled animals
			700	Invertebrates
Proterozoic (Eon)			2500	Metazoans
			3500	Earliest life
Archean (Eon)			4000	Oldest rocks
			4600	Meteorites

Figure 1–1 The surface of the moon viewed from the *Apollo 8* spacecraft, with the Earth rising above the lunar horizon. Courtesy of NASA and USGS

from which life sprang forth. Comets comprising rock debris and ice also plunged into the Earth, releasing tremendous quantities of water vapor and gases. The degassing of these cosmic invaders produced mostly carbon dioxide, ammonia, and methane, major constituents of the early atmosphere, which began to form about 4.4 billion years ago.

Most of the water vapor and gases originated within the Earth itself by volcanic outgassing. Magma contains heavy amounts of volatiles, mostly water and carbon dioxide. Tremendous pressures deep inside the Earth held the volatiles within the magma until it rose to the surface, where the trapped water and gases escaped violently as the magma depressurized. The early volcanoes were extremely explosive because the Earth's interior was hotter and the magma contained more abundant volatiles than is now the case.

The Earth soon acquired a primordial atmosphere composed of carbon dioxide, nitrogen, water vapor, and other gases belched from a profusion of volcanoes. Water vapor so saturated the air that atmospheric pressure was many times greater than it is today. The early atmosphere contained up to 1,000 times the current level of carbon dioxide, which was fortunate because the sun's output was only about 75 percent of its present value and a strong greenhouse effect kept the Earth from freezing. The planet also retained its warmth by a high rotation rate, with days only 14 hours long.

The surface of the Earth was scorching hot and in a constant rage. Winds blew with tornadic force, and fierce dust storms raging across the dry surface blanketed the planet with suspended sediment, much as do Martian dust storms today (Fig. 1–2). Huge lightning bolts darted back and forth, and earth-shaking thunder sent gigantic shock waves reverberating through the air. Volcanoes erupted in one gigantic outburst after another, lighting the sky with white-hot sparks of ash and sending red-hot lava flowing across the land.

The restless Earth rent apart as massive quakes cracked open the thin crust, and huge batches of magma bled through the fissures. Voluminous lava flows flooded the surface, forming flat, featureless plains dotted with towering volcanoes. The intense volcanism also lofted massive quantities of volcanic debris into the atmosphere, giving the sky an eerie red glow. The dust cooled the planet and provided particles around which water vapor coalesced.

With a further drop in atmospheric temperatures, water vapor condensed into heavy clouds that shrouded the planet, completely blocking out the

Figure 1–2 A boulder-strewn field showing rocks embedded in fine sediment from Martian dust storms. Courtesy of NASA

sun and plunging the surface into darkness. As the atmosphere continued to cool, sheets of rain fell from the sky, and deluge upon deluge overflowed the landscape. Raging floods cascaded down steep volcanic slopes and the sides of large meteorite craters, gouging out deep ravines in the rocky plain. Around 4 billion years ago, when the rains ceased and the skies finally cleared, the Earth emerged as a giant blue orb covered by a global ocean nearly 2 miles deep, with scattered chains of volcanic islands.

THE UNIVERSAL SEA

In a remote mountainous area in southwest Greenland (Fig. 1–3), metamorphosed marine sediments of the Isua Formation furnish strong evidence for

Figure 1–3 Location of the Isua Formation in southwestern Greenland, which contains some of the oldest rocks on the Earth.

an early ocean. The continental crust was perhaps only one-tenth of its current size and contained slivers of granite that drifted freely over the Earth's watery face. The Isua rocks originated in volcanic island arcs and therefore lend credence to the idea that plate tectonics operated early in the history of the Earth. The rocks are among the most ancient, dating to about 3.8 billion years, and indicate that the planet had abundant surface water by this time.

Between the end of the great meteorite bombardment and the formation of the first sedimentary rocks about 3.8 billion years ago, large volumes of water flooded the Earth's surface. Seawater probably began salty due to the volcanic outgassing of chlorine and sodium, but the ocean did not reach its present concentration of salts until about 500 million years ago. The salt level has remained generally constant ever since. However, major changes in seawater chemistry often correlated with biological radiation and extinction.

Most of the crust was deeply submerged during the early history of the Earth, as evidenced by an abundance of chert, which is among the hardest minerals and appears to have precipitated from silica-rich water in deep

TABLE 1–2 RADIATION AND EXTINCTION OF SPECIES

Organism	Radiation	Extinction
Mammals	Paleocene	Pleistocene
Reptiles	Permian	Upper Cretaceous
Amphibians	Pennsylvanian	Permian–Triassic
Insects	Upper Paleozoic	
Land plants	Devonian	Permian
Fishes	Devonian	Pennsylvanian
Crinoids	Ordovician	Upper Permian
Trilobites	Cambrian	Carboniferous and Permian
Ammonoids	Devonian	Upper Cretaceous
Nautiloids	Ordovician	Mississippian
Brachiopods	Ordovician	Devonian and Carboniferous
Graptolites	Ordovician	Silurian and Devonian
Foraminiferans	Silurian	Permian and Triassic
Marine invertebrates	Lower Paleozoic	Permian

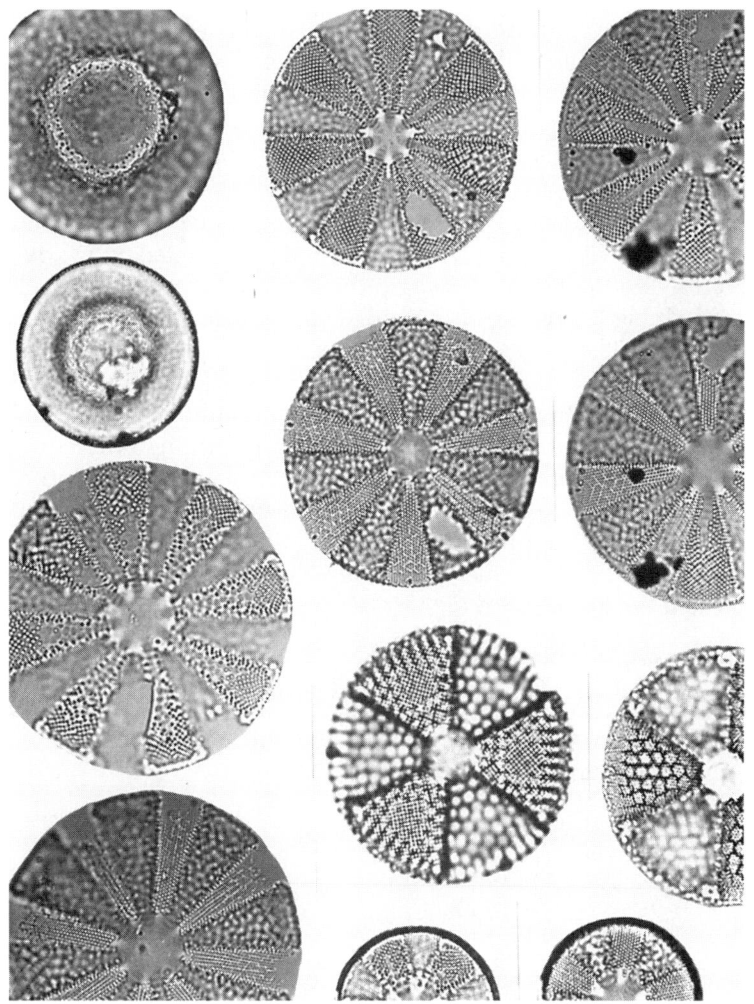

Figure 1–4 Diatoms from the Choptank Formation, Calvert County, Maryland. Photo by G. W. Andrews, courtesy of USGS

oceans. Modern seawater is deficient in silica because such organisms as sponges and diatoms extract it to build their skeletons (Fig. 1–4). Massive deposits of diatomaceous earth, or diatomite, are a tribute to the great success of these organisms during the last 600 million years. Between 10 and 4 million years ago, mats of diatoms spread across vast areas of the eastern tropical Pacific. When the mats sank, they were preserved in the bottom ooze that slowly accumulated during millions of years.

Sulfur was particularly abundant in the early ocean and combined easily with metals like iron to form sulfates. The earliest organisms were sulfur-metabolizing bacteria similar to those living symbiotically in the tissues of tubeworms existing near sulfurous hydrothermal vents on the East Pacific Rise and on a dozen other midocean ridges scattered around the world. Because the atmosphere and ocean lacked significant amounts of oxygen, the bacteria obtained energy by the reduction of sulfate ions. The growth of primitive bacteria was also limited by the amount of organic molecules produced in the ocean.

Iron that was leached from the continents and dissolved in seawater consequently reacted with oxygen in the ocean and precipitated in massive ore deposits on shallow continental margins. Alternating layers of iron-rich and iron-poor sediments gave the ore a banded appearance, called a banded iron formation (BIF). The average ocean temperature was probably warmer than it is today. When warm ocean currents rich in iron and silica flowed toward the polar regions, the suddenly cooled waters could no longer hold minerals in solution; and the minerals precipitated, forming alternating layers because of the difference in settling rates between silica and iron, the heavier of the two minerals. BIF deposits mined extensively throughout the world provide more than 90 percent of the minable iron reserves. In effect, primitive plant life, which generated oxygen by photosynthesis, indirectly created the Earth's iron deposits.

TABLE 1–3 EVOLUTION OF THE BIOSPHERE

Event	Billions of Years Ago	Biological Consequence	Percent Oxygen	Results
Full oxygen conditions	0.4	Fishes, land plants, and animals	100	Approach present biological conditions
Appearance of shelly animals	0.6	Cambrian fauna	10	Burrowing habitat
Metazoans appear	0.7	Ediacarian fauna	7	First metazoan fossils & tracks
Eukaryotic cells appear	1.4	Larger cells	>1	Red beds, multicellular organisms
Blue-green algae	2.0	Algal filaments	1	Oxygen metabolism
Algal precursors	2.8	Stromatolites	<1	Beginning photosynthesis
Origin of life	4.0	Light carbon	0	Evolution of biosphere

Figure 1–5 Archean greenstone belts comprise the ancient cores of the continents.

Oxygen, which currently comprises 21 percent of the atmosphere, was practically nonexistent when life first appeared. Oxygen levels remained low because of the oxidation of dissolved metals in seawater and reduced gases emitted from submarine hydrothermal vents. The seas contained much iron, which reacted with oxygen generated by photosynthesis—a fortunate circumstance, since oxygen was also poisonous to primitive life forms.

About 2 billion years ago, after most of the dissolved iron had been locked up in the sediments, the level of oxygen began to rise and replace carbon dioxide in the ocean and atmosphere. Major plate tectonic cataclysms caused oxygen levels to surge between 2.1 and 1.7 million years ago, and again between 1.1 and 0.7 billion years ago, during the breakup of supercontinents and the formation of new ocean basins.

Greenstone belts comprising ancient metamorphic rocks in the interiors of continents (Fig. 1–5) are among the best evidence for plate tectonics operating early in the Earth's history. Ophiolite complexes in greenstones are slices of ocean floor shoved up on the continents by drifting plates and date as old as 3.6 billion years. Blueschists are metamorphosed rocks of subducted ocean crust thrust onto the continents by plate motions. Pillow lavas, which are tubular bodies of basalt extruded undersea, also appear in the greenstone belts, signifying that the volcanic eruptions took place on the ocean floor.

Plate tectonics has played a prominent role in shaping the Earth virtually from the very beginning of the planet. Continents were adrift from the time they originated, within a few hundred million years after the formation of

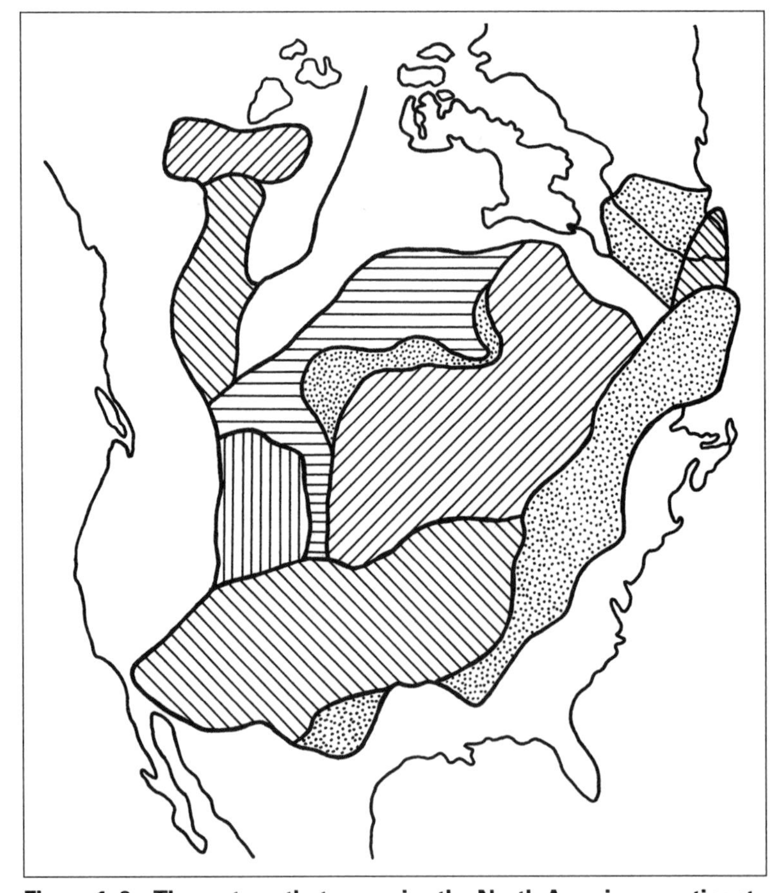

Figure 1–6 The cratons that comprise the North American continent.

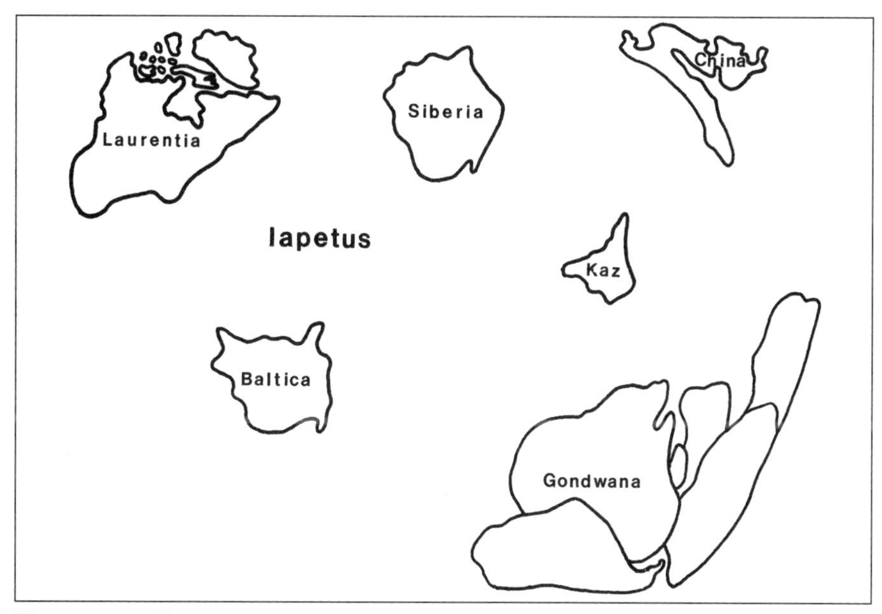

Figure 1–7 About 500 million years ago, the continents surrounded an ancient sea called the Iapetus.

the Earth. This tectonic activity is manifested by 4-billion-year-old Acasta gneiss, a metamorphosed granite, in Canada's Northwest Territories, suggesting that the formation of the crust was well underway by this time. The discovery of this gneiss leaves little doubt that at least small patches of continental crust existed on the Earth's surface during the first half billion years of its existence.

The proto-North American continent called Laurentia assembled from a half dozen major crustal fragments called cratons (Fig. 1–6) some 1.8 billion years ago, making it one of the oldest continents. At Cape Smith on Hudson Bay lies a piece of oceanic crust that was squeezed onto the land during this time, a telltale sign that continents collided and enclosed an ancient sea. Arcs of volcanic rock weave through central and eastern Canada down into the Dakotas. In a region between Canada's Great Bear Lake and the Beaufort Sea lie the roots of an ancient mountain range running through the basement rock, formed by the collision of Laurentia with an unknown landmass more than a billion years ago.

Toward the end of the Precambrian era, some 700 million years ago, all landmasses assembled into a supercontinent centered over the equator. A superocean located approximately in the region of the present Pacific Ocean surrounded the supercontinent. Between 630 and 560 million years ago, the supercontinent rifted apart and 4 or 5 continents rapidly separated. As the continents dispersed and subsided, seas flooded the interiors, creating large continental shelves. Most of the continents huddled around the tropics. This setting heralded an explosion of new life forms in the warm Cambrian seas.

THE IAPETUS SEA

Approximately 500 million years ago, the continents surrounded a large body of water called the Iapetus Sea (Fig. 1–7), which opened as a result of the breakup of the late Precambrian supercontinent. In the Southern Hemisphere, continental motions assembled the present continents of South America, Africa, Australia, Antarctica, and India into the supercontinent Gondwana, named for a geologic province in east-central India.

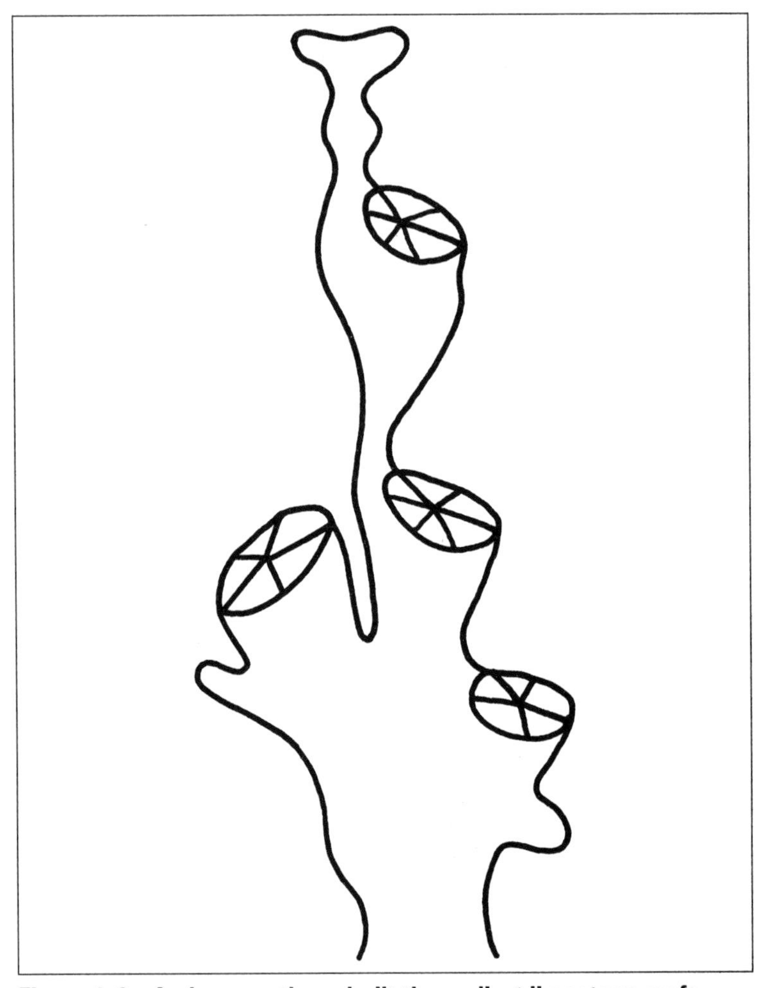

Figure 1–8 Archaeocyathans built the earliest limestone reefs.

Present-day Australia sat at the northern edge of Gondwana on the Antarctic Circle. Fossils of the tropical fern glossopteris, whose leaf impressions look like feathers, appeared in coal beds on the southern continents and India. The plant is absent on the northern continents, which suggests the previous existence of two large continents, one in the Southern Hemisphere and the other in the Northern Hemisphere, separated by a large open sea.

The formation of the Iapetus created extensive inland seas that inundated most of the ancestral North American continent (Laurentia) and the ancient European continent, called Baltica. The Iapetus Sea was similar in size to the present North Atlantic and occupied the same general location. A continuous coastline running from Georgia to Newfoundland between about 570 and 480 million years ago suggests that this ancient east coast faced a wide, deep sea, which stretched at least 1,000 miles across from east to west and bordered a much larger body of water to the south.

Volcanic islands dotted the Iapetus Sea, which resembled the present Pacific Ocean between Southeast Asia and Australia. About 460 million years ago, the shallow waters of the near-shore environment of this ancient sea contained abundant invertebrates, including trilobites—small, oval arthropods that accounted for about 70 percent of all species and a favorite among fossil collectors. Eventually, the trilobites faded, while mollusks and other invertebrates expanded throughout the seas. The vase-shaped archaeocyathans (Fig. 1–8) resembled both sponges and corals and built the

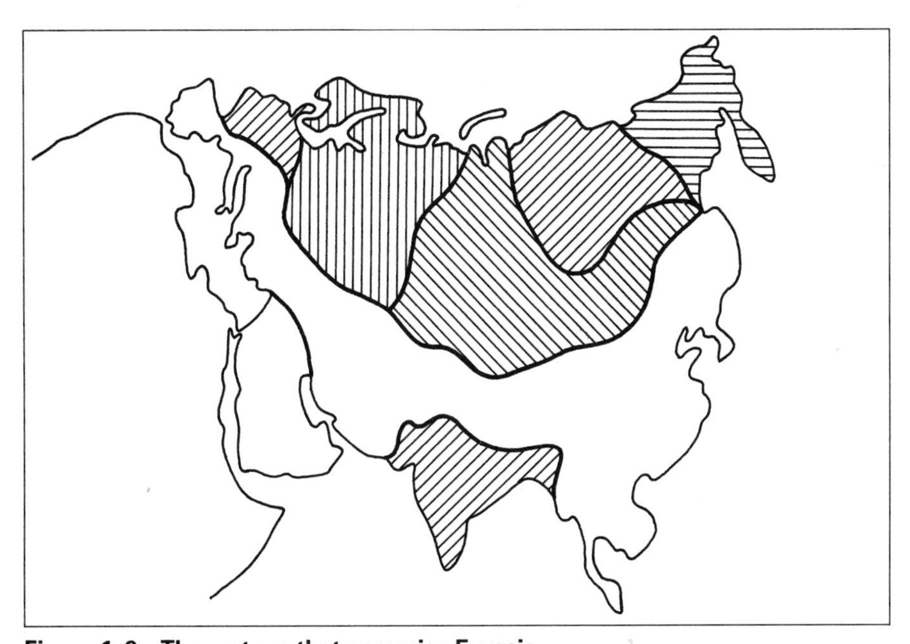

Figure 1–9 The cratons that comprise Eurasia.

earliest limestone reefs, eventually becoming extinct in the Cambrian period.

Between about 420 million and 380 million years ago, Laurentia collided with Baltica, closing off the Iapetus. The collision fused the two continents into the megacontinent Laurasia, named after the Laurentian province of Canada and the Eurasian continent. The Eurasian continent, the largest landmass in the world, formed when some dozen individual continental blocks welded together approximately a half billion years ago (Fig. 1–9).

During the formation of Laurasia, island arcs between the two landmasses were scooped up and plastered against continental edges as the oceanic crustal plate carrying the islands subducted under Baltica. This subduction rafted the islands into collision with the continent and deposited the formerly submerged rocks on the present west coast of Norway. Slices of land called *terranes*, situated in western Europe, drifted into the Iapetus from ancient Africa. Likewise, slivers of crust from Asia traveled across the ancestral Pacific Ocean to form much of western North America.

THE PANTHALASSA SEA

Throughout geologic history, smaller continental blocks collided and merged into larger continents. Millions of years after assembly, the continents rifted apart, and the chasms filled with seawater to form new oceans. However, the regions presently bordering the Pacific Basin apparently have not collided with each other. Rather, the Pacific Ocean is a remnant of an ancient sea called the Panthalassa, which narrowed and widened in response to continental breakup, dispersal, and reconvergence in the area occupied by today's Atlantic Ocean.

So, while oceans have repeatedly opened and closed in the vicinity of the Atlantic Basin, a single ocean has existed continuously at the site of the Pacific Basin. Following the breakup of Pangaea in the early Jurassic period about 170 million years ago, the Pacific plate was hardly larger than the present-day United States. The rest of the ocean floor was comprised of other unknown plates that disappeared as the Pacific plate grew; consequently, no existing oceanic crust is older than Jurassic in age.

Laurentia, comprising the interior of North America, Greenland, and northern Europe, assembled during the collision of several microcontinents beginning about 1.8 billion years ago and evolved during a relatively brief period of only 150 million years. Laurentia continued to grow by garnering bits and pieces of continents and chains of young volcanic islands. About 700 million years ago, Laurentia collided with another large continent on its southern and eastern edges, creating a new supercontinent centered over the equator. A superocean positioned in the approximate location of today's Pacific Ocean surrounded this supercontinent.

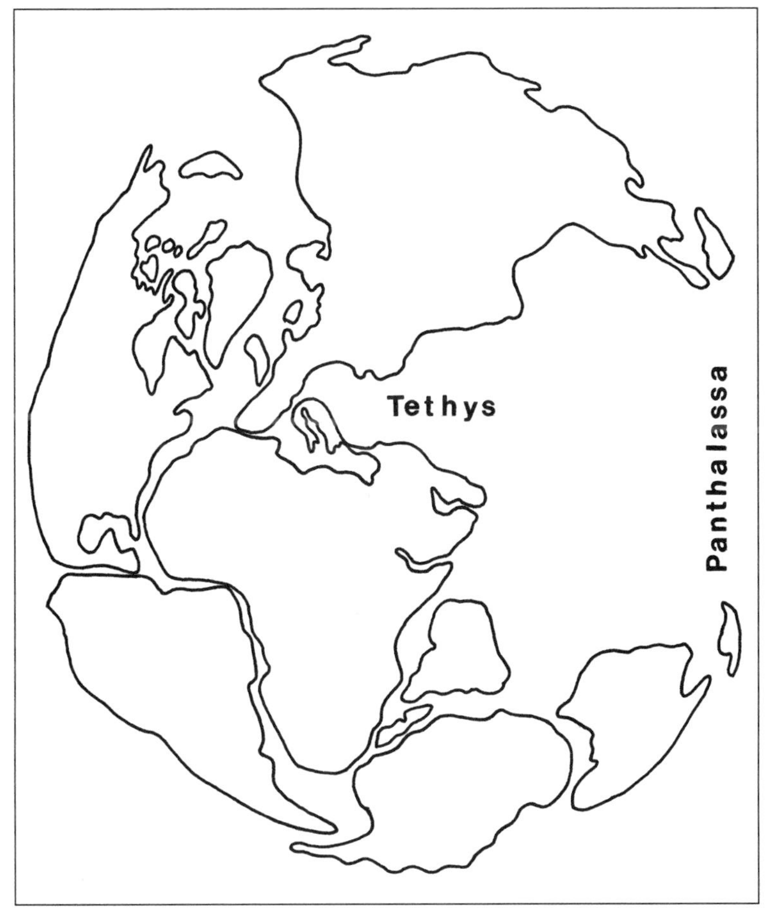

Figure 1–10 The supercontinent Pangaea extended almost from pole to pole.

When Laurentia fused with Baltica to form the megacontinent Laurasia, island arcs in the Panthalassa Sea began colliding with the western margin of the present North America. Erosion leveled the continents, and shallow seas flowed inland, flooding more than half the land surface. The inland seas and wide continental margins, along with a stable environment, enabled marine life to flourish and spread throughout the world.

From 360 million to 270 million years ago, Gondwana and Laurasia converged into Pangaea, which straddled the equator and extended almost from pole to pole (Fig. 1–10). This massive continent reached its peak size about 210 million years ago, with an area of about 80 million square miles or 40 percent of the Earth's total surface area. More than a third of the landmass was covered with water. An almost equal amount of land existed in both hemispheres, whereas today two thirds of the continental landmass is located north of the equator; below the equator the breakdown is 10 percent landmass and 90 percent ocean. A single great ocean stretched uninterrupted across the planet, with the continents huddled to one side of the globe.

The sea level fell substantially after the formation of Pangaea, draining the interiors of the continents and causing the inland seas to retreat. A continuous shallow-water margin ran around the entire perimeter of Pangaea, and no major physical barriers hampered the dispersal of marine life. Moreover, the seas were largely restricted to the ocean basins, leaving the continental shelves mostly exposed.

The continental margins were less extensive and narrower than they are now, due to a drop in sea level as much as 500 feet, which confined marine habitats to the near-shore regions. Consequently, habitat area for shallow-water marine organisms was limited, causing low species diversity. Permian ocean life was sparse, with many immobile animals and few active predators. Ocean temperatures remained cool following a late Permian ice age. Marine invertebrates that managed to escape extinction lived in a narrow margin near the equator. Corals, which require warm, shallow water for survival, were particularly hard hit, as evidenced by the lack of coral reefs at the beginning of the Mesozoic era.

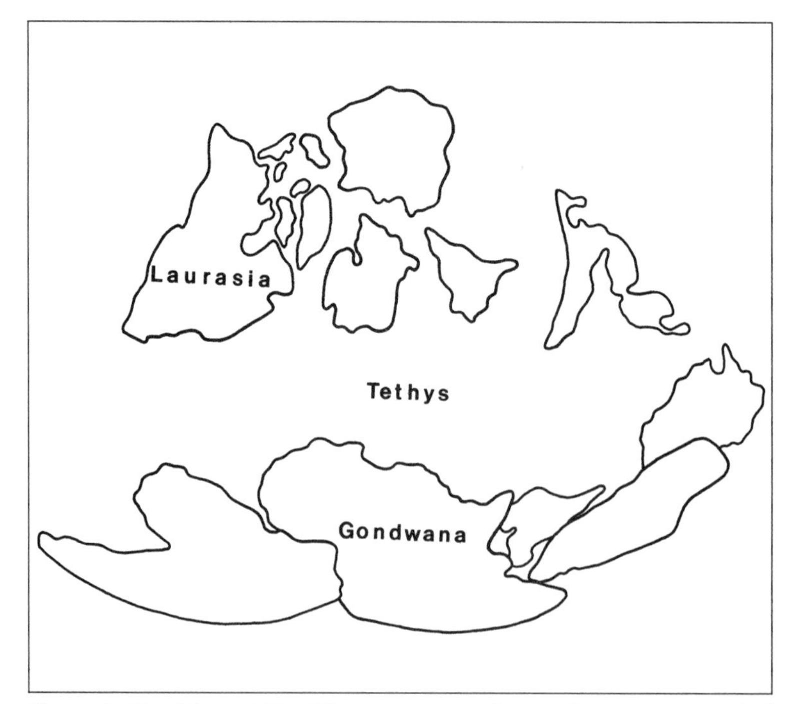

Figure 1–11 About 400 million years ago, the continents surrounded an ancient sea called the Tethys.

THE TETHYS SEA

With Laurasia occupying the Northern Hemisphere and its counterpart Gondwana located in the Southern Hemisphere, the two landmasses were separated by a large shallow equatorial body of water called the Tethys Sea (Fig. 1–11), named for the mother of the seas in Greek mythology. After the assembly of Pangaea, the Tethys became a huge embayment separating the northern and southern arms of Pangaea, which resembled a gigantic letter C straddling the equator.

The Tethys was a broad tropical seaway that extended from western Europe to southeast Asia and harbored diverse and abundant shallow-water marine life. Reef-building was intense in the Tethys Sea, forming thick deposits of limestone and dolomite laid down by prolific lime-secreting organisms. The tropics served as an evolutionary cradle because they had a greater area of shallow seas than other regions, providing an exceptional environment for new organisms to evolve.

During the Mesozoic era, an interior sea flowed into the west-central portions of North America and inundated the area that now comprises eastern Mexico, southern Texas, and Louisiana. Seas also invaded South America, Africa, Asia, and Australia. The continents were flatter, mountain ranges were lower, and sea levels were higher than at present. Thick beds of limestone and dolomite were deposited in the interior seas of Europe and Asia; these rock beds later uplifted to form the Alps and Himalayas.

At the beginning of the Cenozoic era, high sea levels continued to flood continental margins and formed great inland seas, some of which split continents in two. Seas divided North America in the Rocky Mountain and high plains regions, South America was cut in two in the region that later became the Amazon basin, and Eurasia was split by the joining of the Tethys Sea and the newly-formed Arctic Ocean.

The oceans were interconnected in the equatorial regions by the Tethys Sea, providing a unique circumglobal oceanic current system that distributed heat to all parts of the world and maintained an unusually warm climate. The higher sea levels reduced the total land surface to perhaps half its present size. About 80 million years ago, the Western Interior Cretaceous Seaway (Fig. 1–12) was a shallow body of water that divided the North American continent into the western highlands, comprising the newly forming Rocky Mountains and isolated volcanoes, and the eastern uplands, consisting of the Appalachian Mountains.

During the final stages of the Cretaceous period, when waters receded from the land as sea levels dropped, temperatures in the Tethys Sea began to fall. Most warmth-loving species, especially those living in the tropical Tethys Sea, disappeared when the Cretaceous ended. The most temperature sensitive Tethyan fauna suffered the heaviest extinction rates. Species

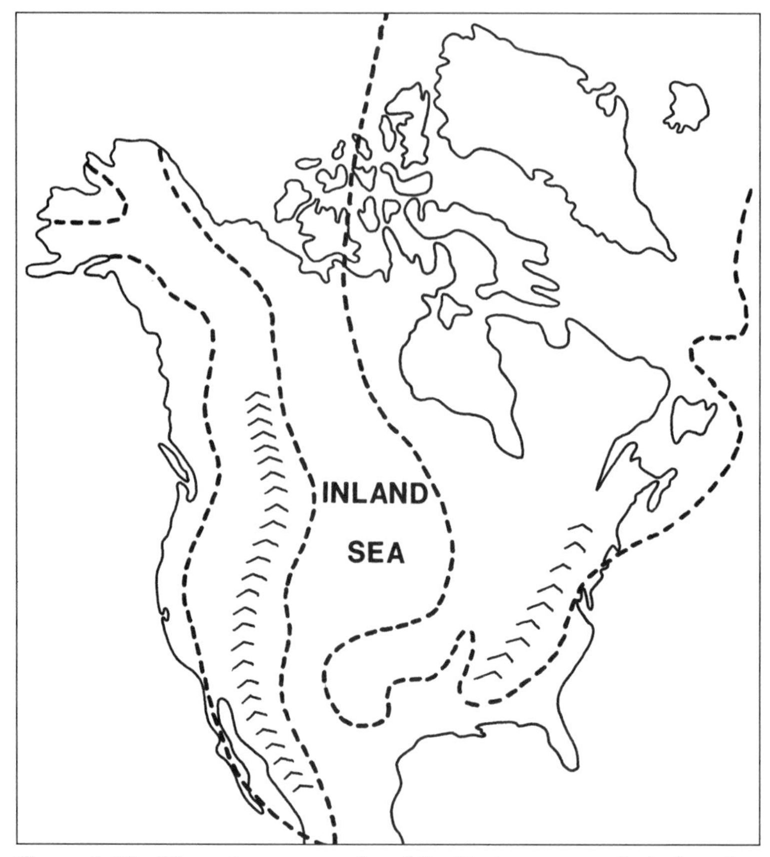

Figure 1–12 The paleogeography of the Cretaceous period, showing the interior sea. Dashed lines indicate ancient landmasses.

that were amazingly successful in the warm waters of the Tethys dramatically declined when ocean temperatures dropped.

Major marine groups that disappeared at the end of the Cretaceous included marine dinosaurs, the ammonoids (ancestors of the nautilus) the rudists (huge coral-shaped clams), and other types of calms and oysters. All the shelled cephalopods were absent in the Cenozoic seas, except the nautilus and shell-less species, including cuttlefish, octopus, and squid. The squid competed directly with fish, which were little affected by the extinction.

Marine species that survived the great extinction were much the same as those of the Mesozoic era. The ocean has a moderating effect on evolutionary processes because it has a longer "memory" of environmental conditions than does land, taking much longer to heat up or cool down. Species that inhabited unstable environments, such as those regions in the higher latitudes, were especially successful. Offshore species fared much better than those in the turbulent inshore waters.

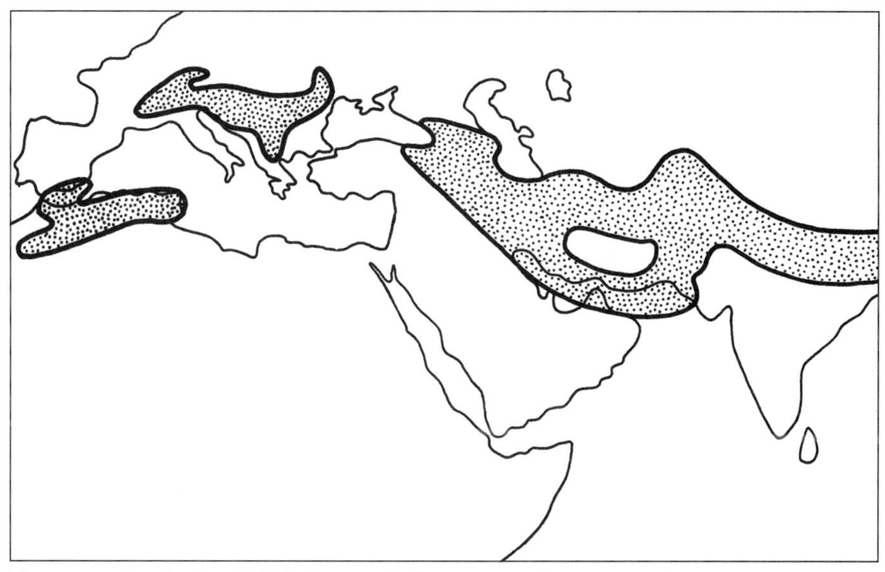

Figure 1–13 Active fold belts result from crustal compression where continental tectonic plates collide, such as the collision of Africa with Eurasia.

About 50 million years ago, the Tethys Sea narrowed as the African and Eurasian continents collided, closing off the sea entirely beginning about 20 million years ago. Thick sediments in the Tethys Sea separating Gondwana and Laurasia buckled and uplifted into mountain belts on the northern and southern flanks (Fig. 1–13). The contact between the continents spurred a major mountain building episode that raised the Alps and other ranges in Europe and squeezed out the Tethys Sea.

When the Tethys linking the Indian and Atlantic oceans closed as Africa rammed into Eurasia, the collision resulted in a long chain of mountains and two major inland seas, the ancestral Mediterranean and a composite of the Black, Caspian, and Aral seas, called the Paratethys, which covered much of what is now Eastern Europe. About 15 million years ago, the Mediterranean separated from the Paratethys, which became a brackish sea. Then, about 6 million years ago, the Mediterranean evaporated, leaving a huge gapping pit whose floor baked in the desert sun. The adjacent Black Sea, a remnant of the ancient Tethys, might have experienced a similar fate.

THE ATLANTIC

About 170 million years ago, a great rift that developed in the present Caribbean Sea began to separate Pangaea into today's continents (Fig. 1–14). The rift sliced northward through the continental crust that connected

North America, northwest Africa, and Eurasia, separating the continents. In the process, this area breached and flooded with seawater, forming the infant North Atlantic. The rifting occurred over a period of several million years along a zone hundreds of miles wide. At about the same time, India nestled between Africa, and Antarctica drifted from Gondwana. Still attached to Australia, Antarctica swung away from Africa toward the southeast, forming the proto-Indian Ocean.

About 50 million years after rifting began, the infant North Atlantic had achieved a depth of 2 miles or greater. It was bisected by an active midocean ridge system that produced new oceanic crust as the plates carrying the surrounding continents separated. Meanwhile, the South Atlantic began to form, opening up like a zipper from south to north. The rift propagated northward at a rate of several inches per year, similar to the separation rate of the two plates carrying South America and Africa. The entire process of opening the South Atlantic took place in a span of just 5 million years.

The breakup of Pangaea compressed the ocean basins, causing a rise in sea levels and a transgression of the seas onto the land. After this breakup,

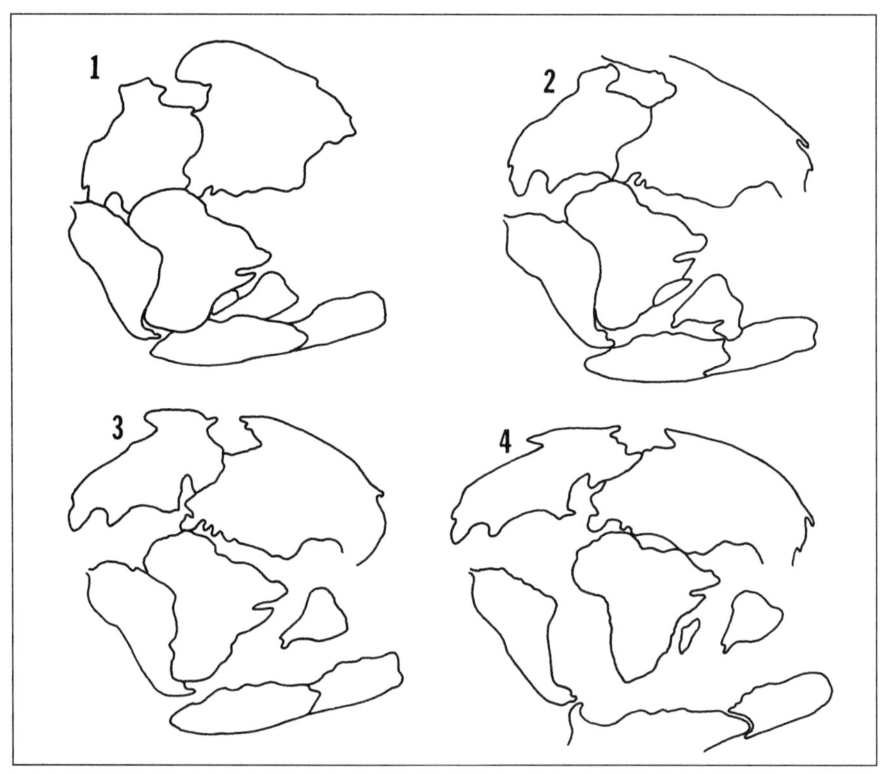

Figure 1–14 The breakup of Pangaea: (1) 225, (2) 180, (3) 135, and (4) 65 million years ago.

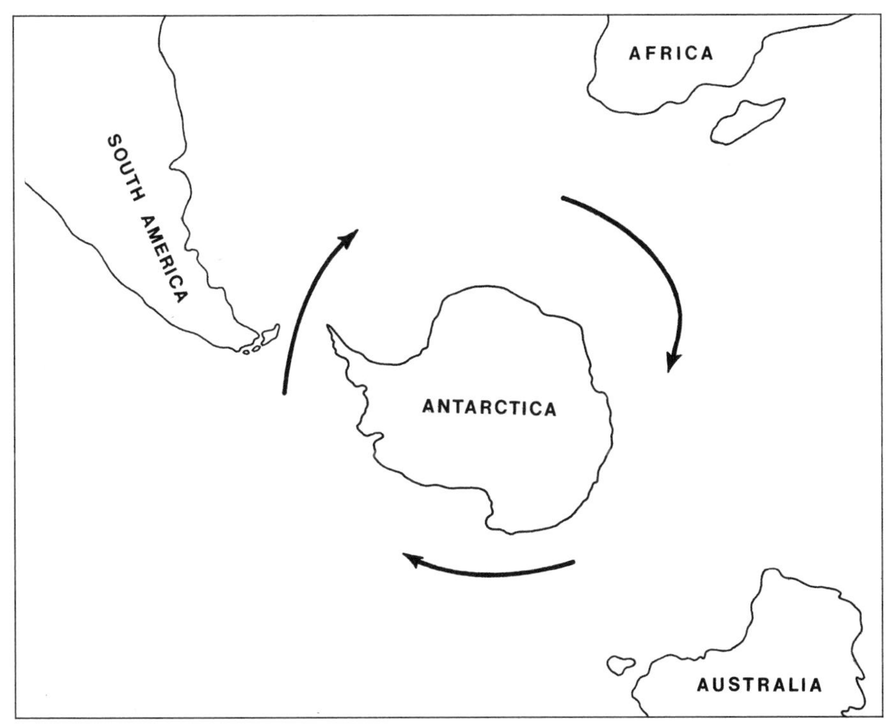

Figure 1–15 The circum-Antarctic current isolates the waters off Antarctica.

the continents drifted apart in spurts instead of a constant speed. The rate of seafloor spreading in the Atlantic matches the rate of plate subduction in the Pacific, where one plate dives under another, forming a deep trench. The subduction of old oceanic crust explains why the ocean floor is no older than 170 million years.

By 80 million years ago, the North Atlantic was a fully developed ocean. Some 20 million years later, the Mid-Atlantic rift progressed into the Arctic Basin, detaching Greenland from Europe. North America was no longer connected to Europe except for a land bridge across Greenland that continued to allow the migration of species between the two continents. The strait between Alaska and Asia narrowed, creating the nearly landlocked Arctic Ocean.

The South Atlantic continued to widen, with more than 1,500 miles of ocean separating South America and Africa. Africa moved northward, leaving Antarctica (still joined to Australia), behind, and began to close the Tethys Sea. In the early Tertiary, Antarctica and Australia broke away from South America and moved eastward. When the two continents rifted apart, Antarctica moved toward the South Pole, while Australia continued moving northeastward.

When Antarctica separated from South America and Australia and drifted over the South Pole some 40 million years ago, the polar vortex formed a circumpolar Antarctic ocean current (Fig. 1–15). This current isolated the frozen continent, preventing it from receiving warm poleward flowing waters from the tropics. Deprived of warmth, Antarctica became a frozen wasteland. During this time, warm salt water filled the ocean depths, while cooler water covered the upper layers.

Because of high evaporation rates and low rainfall, warm water in the Tethys Sea became top-heavy with salt and sank to the sea's bottom. Meanwhile, ancient Antarctica, whose climate was warmer than it is today, generated cool water that filled the sea's upper layers, causing the deep ocean to circulate from the tropics to the poles, just the opposite of today's patterns. About 28 million years ago, Africa collided with Eurasia and blocked warm water from flowing to the poles, thereby allowing a major ice sheet to form on Antarctica. Ice flowing into the surrounding sea cooled the surface waters, which sank to the ocean depths and flowed toward the equator, generating the present ocean circulation system.

The early Tertiary coincided with changes in the deep-ocean circulation. Land gathering in one area affects the shapes of ocean basins. The ocean bottom influences how much heat ocean currents carry from the tropics to the poles. The change in ocean circulation eliminated many species of marine life on the European continent, which flooded with shallow seas.

When Greenland separated from North America and Eurasia, beginning about 57 million years ago, it opened the North Atlantic. The separation of Greenland from Europe might have drained frigid arctic waters into the North Atlantic, significantly lowering its temperature. The climate grew much colder and the seas withdrew from the land as the ocean fell by about 1,000 feet to perhaps its lowest level in the last several hundred million years; it remained depressed for the next 5 million years. The drop in sea level also coincided with the accumulation of massive ice sheets atop Antarctica.

Greenland was largely ice-free until about 4 million years ago, when a sheet of ice up to 2 miles thick buried this large island. Alaska connected with eastern Siberia and closed off the Arctic Basin from warm water currents originating in the tropics, resulting in the formation of pack ice in the Arctic Ocean.

About 3 million years ago, the Isthmus of Panama separating North and South America uplifted as oceanic plates collided. The barrier created by the land bridge isolated Atlantic and Pacific species, and extinctions impoverished the once rich fauna of the western Atlantic. The new landform halted the flow of cold water currents from the Atlantic into the Pacific, that, along with the closing of the Arctic Ocean from warm Pacific currents, might have initiated the Pleistocene ice ages, when massive glaciers swept out of the polar regions and buried the northern lands.

2

MARINE EXPLORATION

Early geologists thought the ocean floor was a barren desert, covered by thick, muddy sediments washed off the land and by debris of dead marine organisms raining down from above. After billions of years, it was assumed that the sediments had accumulated to a depth of several miles, making the deep waters of the ocean a vast, featureless plain, unbroken by ridges or valleys and interspersed with a few scattered volcanic islands.

As technology improved, the view of the seabed grew much more accurate and complex, revealing midocean ridges grander than terrestrial mountain ranges and chasms deeper than any canyon on the continents. The midocean ridges, with their vigorous volcanic activity, appeared to generate new oceanic crust. The deep-sea trenches, with their extensive earthquake activity, seemed to devour old oceanic crust. Strange sea creatures were found on the deep seafloor, where previously no life was thought to exist. Indeed, the ocean floor was much more complicated than ever imagined.

EXPLORING THE OCEAN FLOOR

In the mid-1800s, depth soundings of the ocean floor were taken in preparation for laying the first transcontinental telegraph cable linking the

United States with Europe. The depth recordings indicated hills, valleys, and a middle Atlantic rise (named Telegraph Plateau) where the ocean was supposed to be the deepest. Sometimes, sections of the telegraph cable became buried under submarine slides and had to be brought to the surface for repair.

In 1874, the British cable-laying ship HMS *Faraday* was attempting to mend a broken telegraph cable in the North Atlantic. The cable rested on the ocean floor at a depth of 2.5 miles, where it passed over a large rise. While grappling for the cable, the claws of the grapnel snagged on a rock. When the grapnel was finally freed and brought to the surface, clutched in one of its claws was a large chunk of black basalt, a common volcanic rock. What made this discovery astonishing was that volcanoes were not supposed to be on the Atlantic Ocean floor.

The British corvette HMS *Challenger*, the first fully equipped oceanographic research vessel, was commissioned in 1872 to explore the world's oceans. The crew took depth soundings, water samples, and temperature readings, and dredged bottom sediments for evidence of animal life living on the deep seafloor. The *Challenger*'s nets hauled up large numbers of deep-sea and bottom-dwelling animals, many from the deepest trenches. The catch included some of the strangest creatures scientists had yet found. Many species were unknown to science, and some were thought to have long gone extinct.

During nearly 4 years of exploration, the *Challenger* charted 140 square miles of ocean bottom and sounded every ocean except the Arctic. The deepest sounding was taken off the Mariana Islands in the western Pacific. While recovering samples in the deep waters off the Marianas, the research vessel encountered a deep trough known as the Mariana Trench. This trench forms a long line northward from the Island of Guam and is the lowest place on Earth, reaching a depth of nearly 7 miles below sea level.

While dredging the deep ocean floor in the Pacific, the *Challenger* recovered rocks resembling dense lumps of coal. Mistaken for fossils or meteorites, the rocks were put on display in the British Museum as geological oddities. Almost a century later, further analysis showed the true value of the dark, potato-size clumps. The nodules contained large quantities of valuable metals, including manganese, copper, nickel, cobalt, and zinc. Scientists realized that the world's largest reserve of manganese nodules lay on the bottom of the North Pacific, about 16,000 feet below the surface. Fields thousands of miles long contained nodules estimated at 10 billion tons.

Other valuable minerals were found on the deep-sea floor. In 1978, the French research submersible *Cyana* discovered unusual lava formations and mineral deposits on the seabed in the eastern Pacific, more than 1.5 miles deep. These deposits were sulfide ores in 30-foot-high mounds of porous gray and brown material. The massive sulfide deposits contained abundant iron, copper, and zinc. The French research vessel *Sonne* found another sulfide ore field nearly 2,000 miles long on the floor of the East Pacific. The sediments contained as much as 40 percent zinc along with deposits of other metals, some in greater concentrations than their land-based counterparts.

Research vessels discovered valuable sediments more than 7,000 feet deep on the bed of the Red Sea between Sudan and Saudi Arabia. The largest deposit was in an area 3.5 miles wide known as the Atlantis II Deep, named for the research vessel that discovered it. The rich bottom ooze was estimated to contain about 2 million tons of zinc, 400,000 tons of copper, 9,000 tons of silver, and 80 tons of gold. The sea undoubtedly provided unheard of mineral riches.

In the early 1970s, knowledge of the seafloor and the capacity to explore it were still rudimentary. Shipboard sonar was inadequate for mapping the rugged topography of the midocean ridges. The imagery improved substantially when sonar devices were mounted on a vehicle and towed at a considerable depth behind a ship. A system called SeaBeam made high-resolution sonar maps of the midocean ridge crests. Its sonar covered a broad swath of seafloor, allowing a ship to map an entire area by tracking back and forth in well-spaced lines.

Cameras were also mounted on sleds (Fig. 2–1) and pulled through elaborate obstacle courses in the dark abyss, but the instruments were

Figure 2–1 A deep-sea camera and color video system used to photograph sulfide ore deposits on the seafloor. Photo by Hank Chezar, courtesy of USGS

damaged or lost at an alarming rate. A massive camera vehicle called Angus weighed 1.5 tons, allowing it to be towed almost directly beneath the ship for better navigational control. The most sophisticated device, called Deep Tow, carried sonar, television cameras, and sensors for measuring temperature, pressure, and electricity. During operation over the East Pacific Rise off the coast of Ecuador, the camera sled "flew" into a hot plume of water. Upon further exploration, photographs taken by Angus revealed a lava field scattered with large white clams.

The submersible *Alvin* was sent down to investigate this phenomenon. It discovered an oasis of hydrothermal vents (Figs. 2–a & 2–b) and exotic deep-sea creatures 1.5 miles below sea level. At the base of jagged basalt cliffs was evidence of active lava flows, including fields strewn with pillow

Figure 2–2a The deep submersible *Alvin* at the port of Wood's Hole, Massachusetts. Photo by R. A. Wahl, courtesy of U.S. Navy

Figure 2–2b A hydrothermal vent with sulfide-laden hot water pouring out into cold seawater on the ocean floor. The photograph is taken from *Alvin,* whose claw holds a temperature probe. Photo by N. P. Edgar, courtesy of USGS

lavas. Unusual chimneys called black smokers spewed out hot water blackened with sulfide minerals. Others called white smokers, ejected hot water that was milky white. Species previously unknown to science lived in total darkness among the hydrothermal vents. Tubeworms up to 10 feet tall swayed in the hydrothermal currents. Giant crabs scampered blindly across the volcanic terrain. Huge clams up to a foot long and clusters of mussels formed large communities around the vents.

SURVEYING THE SEABED

The more scientists probed the ocean floor, the more complex they learned it to be. The ocean covers about 70 percent of the Earth's surface to an average depth of over 2 miles. It is shallowest in the Atlantic Basin and deepest in the Pacific Basin. If Mount Everest, the world's tallest mountain, were placed in the deepest part of the Pacific Basin, the water would still rise over a mile above it's peak. Yet in relation to the overall size of the Earth, the ocean is merely a thin layer of water like the outer skin of a onion.

Early methods of sampling the seabed included dragging a dredge behind a ship to scoop up the bottom sediments, or using a snapper (Fig. 2–3)

Figure 2–3 A snapper sampling instrument, whose jaws close when striking the ocean bottom. Photo by K. O. Emery, courtesy of USGS

whose jaws closed when the instrument struck the bottom. But these techniques only sampled the topmost layers, which were not recovered in the order of their original deposition. In the early 1940s, a piston corer was developed, which when dropped to the seabed retrieved a vertical section of the ocean floor intact. The corer consisted of a long barrow that plunged into the bottom mud under its own weight. A piston firing upward from the lower end of the barrow sucked up sediments into a pipe, and these core samples were then brought to the surface (Fig. 2–4).

The bottom of the ocean was at first thought to contain sediments washed off the continents, forming deposits several miles thick after billions of years of accumulation. However, core drilling at several sites revealed that the oldest sediments were less than 200 million years old. The sediments were measured with an undersea device that used seismic waves similar to sound waves to locate sedimentary structures.

An ocean-bottom seismograph dropped to the seafloor (Fig. 2–5) recorded microearthequakes in the Earth's submarine crust and rose automatically to the surface for recovery. Seismic instruments towed behind ships also detected geologic structures deep within the sub-oceanic crust. These surveys provided important information about the ocean floor that could not be obtained by direct means, and revealed that instead of miles of silt and mud, the oceanic crust contained only a few thousand feet of sediment.

During the height of the cold war in the late 1950s, American and Russian oceanographic vessels mapped the ocean floor so that ballistic missile submarines would be able to navigate in deep water without grounding on uncharted seamounts. When Russian aircraft shot down a Korean airliner over Sakhalin Island on August 30, 1983, killing all 269 passengers and

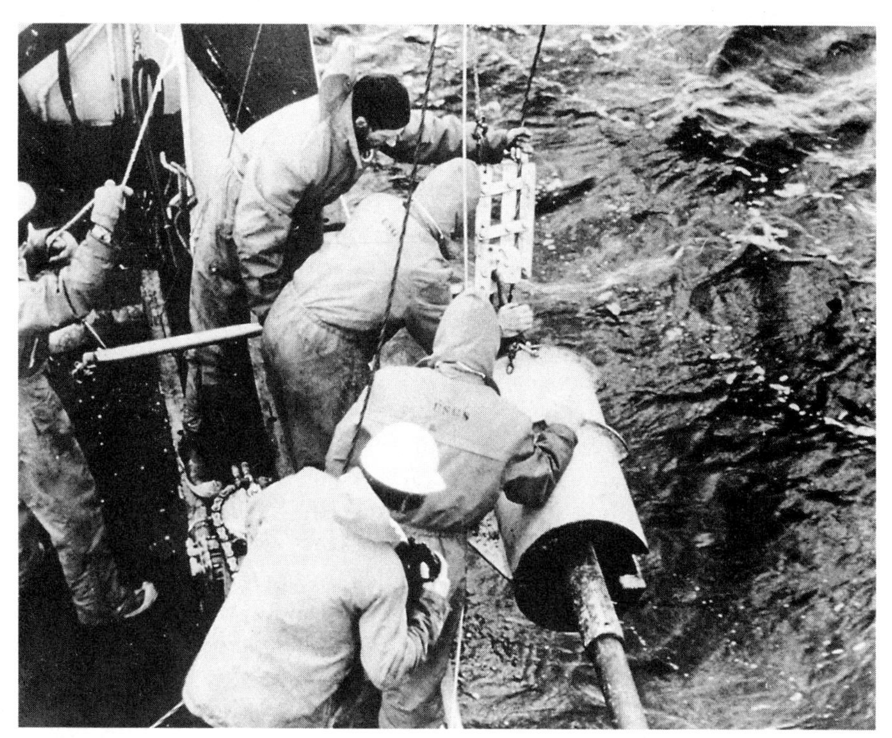

Figure 2–4 Piston coring in the Gulf of Alaska. Photo by P. R. Carlson, courtesy of USGS

Figure 2–5 An ocean bottom seismograph provides direct observations of earthquakes on midocean ridges. Courtesy of USGS

crew, a search for the downed aircraft was conducted using the unmanned submersible *Deep Drone* (Fig. 2–6) operated by the U.S. Navy.

Sonar depth-ranging was another important tool for mapping undersea terrain. SeaMarc, a side-looking sonar system towed in a "fish" about 1,000 feet above the ocean floor, provided a sonar image of the ocean bottom by bouncing sound waves off the seabed (Fig. 2–7). As ships traversed the Atlantic Ocean, onboard sonographs painted a remarkable picture of the ocean floor. Lying 2.5 miles deep in the middle of the Atlantic Ocean was a huge submarine mountain range, surpassing in scale the Alps and the Himalayas. The range ran down the middle of the ocean, weaving halfway between the continents that surrounded the Atlantic Basin. This massive ridge was the site of volcanic activity so intense that it seemed as though the Earth's insides were coming out.

The midocean ridges were found to be a string of seamounts in a region where it was assumed that the deep seafloor should have been flat and barren. With more detailed mapping of the ocean floor, scientists found that the Mid-Atlantic Ridge was the most peculiar mountain range yet

Figure 2–6 The unmanned submersible *Deep Drone* being launched to search for the wreckage of Korean Air Lines Flight 007, shot down near Sakhalin Island on August 30, 1983, by Russian aircraft. Photo by F. Barbante, courtesy of U.S. Navy

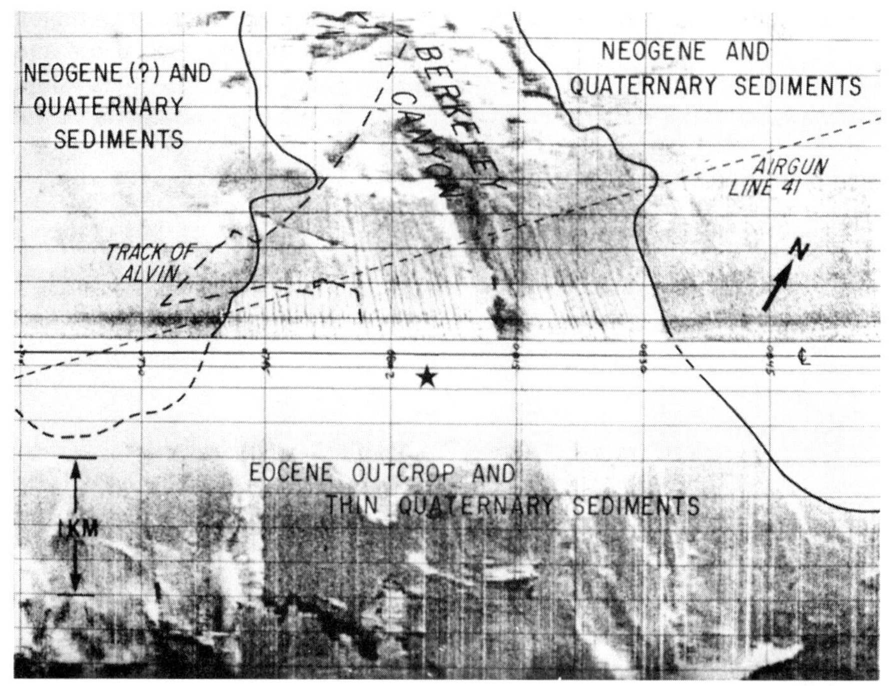

Figure 2–7 Sonograph of the lower continental slope off the Atlantic coast from SeaMarc. Photo by N. P. Edgar, courtesy of USGS

discovered. The ridge crest was 10,000 feet above the ocean floor, with a deep trough running through the middle like a giant crack in the Earth's crust. It was 4 miles deep in places, or four times deeper than the Grand Canyon, and up to 15 miles wide, making it the grandest canyon on Earth.

Undersea survey has shown that the submerged mountains and undersea ridges form a continuous chain 45,000 miles long, several hundred miles wide, and up to 10,000 feet high that winds around the globe like the stitching on a baseball. Although the midocean ridge system lies deep beneath the sea, it is easily the most dominant feature on the face of the planet, extending over an area greater than that covered by all major terrestrial mountain ranges combined.

When advanced instrumentation was developed, the view of the seafloor came into better focus. The ocean floor proved to be far more active and younger than previously imagined. Additional surveys conducted across the extensive undersea mountain ranges included rock sampling, sonar depth-finding, thermal measurements, magnetic readings, and seismic surveys. The resulting data suggested that the oceanic crust was spreading outward at the midocean ridge. Magma rising from the mantle erupted onto the ocean floor, adding new oceanic crust to the ridge crest as both sides pulled apart.

Temperature surveys showed anomalous amounts of heat seeping out of the Earth in the mountainous regions of the middle Atlantic. It was as though magma were bleeding from the mantle through cracks in the oceanic crust. Volcanic activity in the ridges suggested that new material was being added to the seafloor. This activity appeared more intense in the Atlantic Ocean, where the midocean ridge is steeper and more jagged, than in the Pacific or the Indian oceans, where branches of oceanic ridges were overridden by continents.

Deep-sea trenches off continental margins and volcanic island arcs were initially thought to have been created by the tremendous weight of sediments washed off the continents and pulled down into the mantle by a dense underlying material. The downward pull on the sediments formed vast bulges in the ocean floor called geosynclines. However, gravity surveys conducted over the trenches indicated that the pull of gravity was much too weak to account for the sagging of the seafloor.

The trenches were also found to be sites of almost continuous earthquake activity deep in the bowels of the Earth. The deep-seated earthquakes acted like beacons marking the boundaries of a large slab of crust descending into the mantle. The unusual activity of the trenches suggested that they were sites where old oceanic crust subducted into the Earth's interior. Perhaps here at last was the engine that drove the continents around the surface of the Earth.

GEOLOGIC OBSERVATIONS

Observations of these and other fascinating geologic features on the ocean floor led to the development of the seafloor spreading theory. The hypothesis described the creation and destruction of the ocean floor at specific regions around the world. The seafloor spreading theory resolved many problems connected with the mysterious characteristics on the seafloor, including the midocean ridges, the relatively young ages of rocks in the oceanic crust, and the formation of island arcs. But more importantly, here at last was the long-sought mechanism for continental drift. The continents do not plow through the ocean crust like icebreakers slicing through frozen seas, as previously thought, but instead ride above a pliable mantle like ships caught in mobile icefloes.

Exploration of the ocean floor brought a new understanding of the forces that shaped the planet. After overwhelming geological and geophysical evidence was collected from the floor of the ocean in support of the theory of continental drift, geologists finally abandoned the archaic thinking of the past century. By the late 1960s, most geologists in the Northern Hemisphere, who had long rejected the theory, finally joined their southern colleagues, who had been for some time convinced of the reality of conti-

TABLE 2–1 CONTINENTAL DRIFT

Geologic division (millions of years)		Gondwana	Laurasia
Quaternary	3		Opening of Gulf of California
Pliocene	11	Spreading begins near Galapagos Islands	Spreading changes directions in eastern Pacific
		Opening of the Gulf of Aden	
			Birth of Iceland
Miocene	26		
		Opening of Red Sea	
Oligocene	37		
		Collision of India with Eurasia	Spreading begins in Arctic Basin
Eocene	54		Separation of Greenland from Norway
Paleocene	65	Seperation of Australia from Antarctica	
		Seperation of New Zealand from Antarctica	Opening of the Labrador Sea
			Opening of the Bay of Biscay
		Separation of Africa from Madagascar and South America	Major rifting of North America from Eurasia
Cretaceous	135		
Jurassic	180	Separation of Africa from India, Australia, New Zealand, and Antarctica	Seperation of North America from Africa begins
Triassic	250	Assembly of all continents into the supercontinent Pangaea	

nental drift because of overwhelming evidence in South America and the opposing African continent.

The discovery of many mysteries on the seabed, including spreading ridges and deep-sea trenches, led geologists to develop an entirely new way of looking at the Earth: the theory of plate tectonics (Fig. 2–8). Tectonics (from the Greek word *tekton*, meaning "to build") is the geologic process responsible for all features on the Earth's surface. The theory incorporated the process of seafloor spreading and continental drift into a comprehensive model. Therefore, all aspects of the Earth's history and structure could be unified by the revolutionary concept of movable plates. Well-defined earthquake zones marked the boundaries of the plates, and analysis of earthquakes around the Pacific Ocean revealed a consistent direction of crustal movement.

The Atlantic Ocean is bisected by the Mid-Atlantic Ridge, which manufactures new oceanic crust as the continents surrounding the Atlantic Basin spread apart. The Mid-Atlantic Ridge is the center of intense seismic and volcanic activity and the focus of high heat flow from the Earth's interior. Molten magma originating from the mantle rises through the lithosphere and erupts on the ocean floor, adding new oceanic crust to both sides of the ridge crest.

As the Atlantic Basin widens, the surrounding continents separate at a rate of about 1 inch per year. In response to the widening seafloor in the Atlantic and the separation of continents around the Atlantic Basin, the Pacific Basin shrinks at a corresponding rate. The Pacific is ringed by subduction zones that destroy old oceanic crust in deep-sea trenches (Fig.

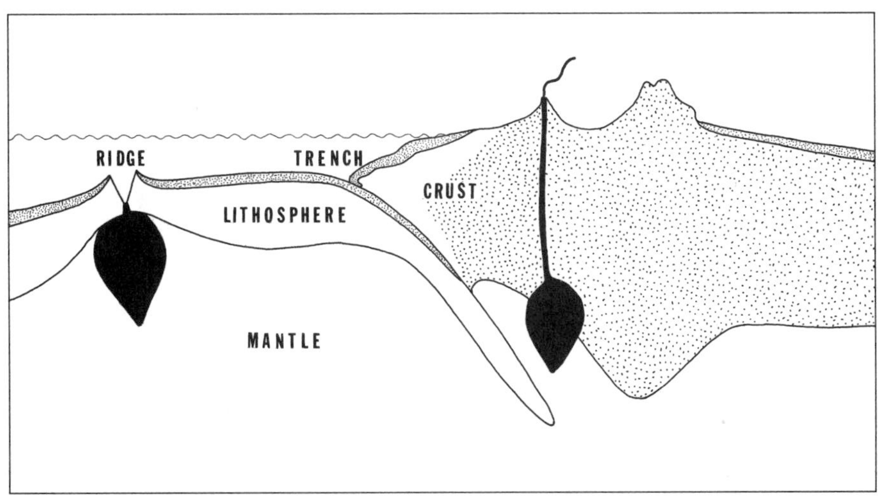

Figure 2–8 The plate tectonics model. New oceanic crust is generated at spreading ridges and old oceanic crust is destroyed in subduction zones, which moves the continents around the face of the Earth.

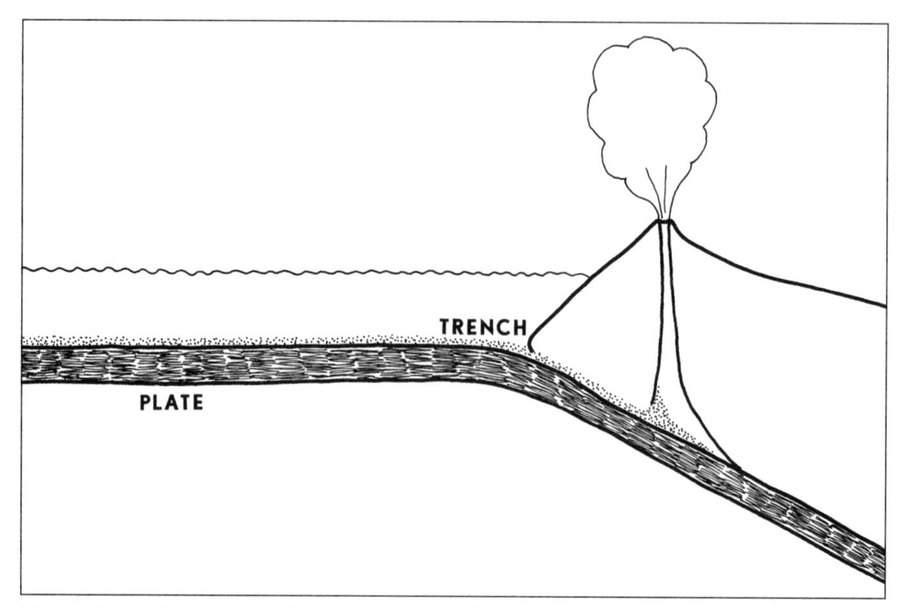

Figure 2–9 The subduction of the ocean floor provides new molten magma for volcanoes that fringe the deep-sea trenches.

2–9). Spreading ridges in the Pacific are also much more active than those in the Atlantic. These features on the ocean floor are responsible for most of the geologic activity that surrounds the Pacific Ocean.

OCEAN DRILLING

The oceans have an average depth of over 2 miles and are blanketed by layers of thick sediments. To correctly date these sediments, they had to be recovered in the order they were laid down; thus, dredging techniques were of little use. Fortunately, a technique known as seafloor coring was developed, enabling scientists to take accurate sediment samples. A hollow pipe is drilled into the sediments and a long cylindrical sample is brought to the surface. Early attempts at coring in deep water, however, only penetrated a few feet into the upper sediments of the ocean floor.

In the mid-1960s the National Science Foundation sponsored a deep-sea drilling program called Project Mohole. The moho, named after the Yugoslav seismologist Andrija Mohorovicic, is the point of contact between the Earth's crust and mantle. The crust is the thinnest in the oceans, measuring only about 3 to 5 miles thick. Scientists hoped that the moho would provide new clues about the origin, age, and composition of the Earth's interior, which land-based drilling could not obtain. Unfortunately, the task of drilling through miles of oceanic crust in waters as much as 2 or more miles deep became expensive and time-consuming.

In 1968, the British research vessel *Glomar Challenger* was commissioned for the Deep Sea Drilling Project, a consortium of American oceanographic institutions. The project's objective was to drill a large number of shallow holes in widely scattered parts of the ocean floor in an attempt to prove the theory of seafloor spreading. A similar deep-sea drillship called the *Glomar Pacific* (Figs. 2–10a&b) was the first to begin drilling on the Atlantic outer continental shelf and slope of the United States. Both ships were designed with a 140-foot drilling derrick amidships; computerized thrusters located fore and aft maintained station over the drill hole even in rough seas.

A string of drill pipe dangled as much as 4 miles beneath the ship, with the drill bit cutting through the sediment by the force of its own weight. The core, which is a cylindrical vertical section of rock, was retrieved through the drill stem by a removable inner barrel, allowing the drill bit to remain in the hole. When the drill bit dulled, it and the drill pipe had to be brought back up to the surface for replacement. The drill string was then lowered back over the drill hole and a special funnel-like apparatus guided the drill bit into the hole.

The primary purpose of the international Ocean Drilling Program (ODP) and the Joint Oceanographic Investigation for Deep Earth Sampling

Figure 2–10a The *Glomar Pacific* drilling on the Atlantic outer continental shelf and slope of the United States. Courtesy of USGS

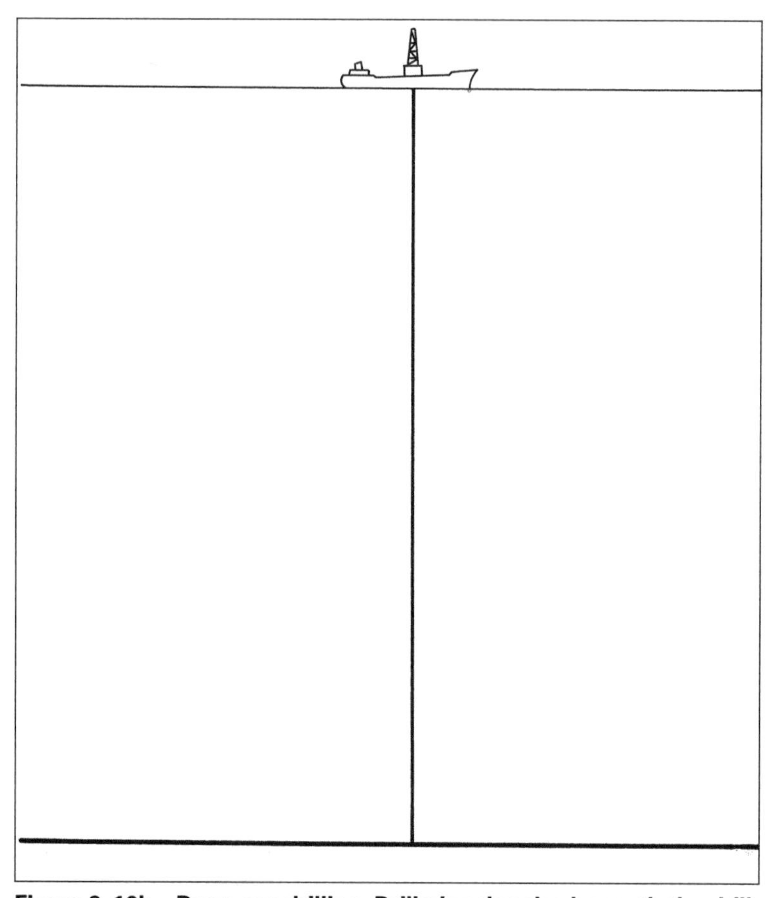

Figure 2–10b Deep-sea drilling. Drill pipe dangles beneath the drillship and cuts through the bottom sediments by its own weight.

(JOIDES) was to take rotary core samples of the ocean floor at hundreds of sites around the world. However, special precautions were taken not to drill in potentially productive oil fields, where drilling might cause blowouts that would result in hazardous oil spills. Just the opposite occurred on the south flank of the Costa Rica Rift east of the Galapagos Islands in 1979 when the *Challenger* drilled a hole into the crust. Instead of blowing out hot water, which is often the case, the well sucked in a powerful, steady stream of seawater. The suction resulted from the downward convection of circulating water within the oceanic crust as it descended toward a magma chamber, acquiring heat during hydrothermal activity.

The deepest hole was bored into the ocean floor in the eastern Pacific near the Galapagos Islands by the drillship JOIDES *Resolution*. The purpose was to sample a section of the entire oceanic crust from top to bottom in an area where the crust was though to be thinnest. During a 14-year period beginning in 1979, the ship made 7 trips to the drill site to deepen the hole, with each session lasting up to 2 months. On the sixth trip, the ship first had to recover drill pipe lost in the hole during the previous effort. When this task was accomplished, the hole was extended further to a depth of more than 6,500 feet beneath the seabed. In January 1993, the *Resolution* returned again to deepen the hole another 370 feet only to lose the drill bit. This mishap forced the crew to abandon the drill hole perhaps only a few hundred feet short of their goal.

Taking a shortcut to the bottom of the ocean crust, ODP scientists found a site where the lower crust is uncovered along the Atlantis II Fracture Zone in the Indian Ocean, which is part of the midocean ridge that forms the boundary between the African and Antarctic tectonic plates. Running down the middle of the ridge is a feature called a spreading center, which periodically breaks apart, leaving a gap that fills with molten magma. As the magma cools and hardens, the rock forms new oceanic crust that joins to the ends of the plates.

The structure of a spreading center resembles steps in a staircase, with short, straight segments roughly parallel to each other (see Fig. 3–2). The fracture zones are valleys that connect adjacent segments like the vertical jumps between the steps. When the scientists drilled through the valley floor of the fracture zone, they recovered coarsely crystalline rocks called gabbros, which are known to make up the lower segment of the oceanic crust.

After recovering and dating cores from several midocean ridges, the *Challenger* made a truly remarkable discovery. The farther away from the deep-sea ridges the ship drilled, the thicker and older the sediments became. But even more surprising was that the thickest and oldest sediments were not billions of years old as expected, but were in fact younger than 200 million years. Near the continental shelves, where thick layers of sediment form flat abyssal plains, the drill cores revealed thin beds of calcium carbonate just above hard volcanic rock that was buried under thousands of feet of red clay and other sediments. The discovery of abyssal red clay, whose color signifies a terrestrial origin, provided additional evidence for seafloor spreading.

The deepest abysses in the world are adjacent to continental margins, the actual boundaries of continents, where the oceanic lithosphere is the oldest. The calcium carbonate layer located by the *Challenger* was about 4 miles deep, far below the depth where the crush of cold water dissolves calcium carbonate. Well protected from the corrosive effect of seawater by the overlying sediments, the calcium carbonate originating in shallower water near midocean ridges was somehow transported to the edges of the continents.

The floor of the Atlantic conveys lithosphere, the rigid layer of the upper mantle, away from its point of origin at the Mid-Atlantic Ridge. The ocean floor at the crest of the midocean ridge consists mostly of basalt, a black volcanic rock. Continuing away from the crest, the bare rock is blanketed by an increasing thickness of sediments, composed mostly of red clay from detritus material washed off the continents and from windblown desert sediment that has landed in the sea. Some large sandstorms over the Sahara Desert blow dust so high into the atmosphere that prevailing air currents carry the dust all the way across the Atlantic Ocean to South America (Fig. 2–11).

Near the ridge crest, the sediments are predominantly composed of calcareous ooze built up by a rain of decomposed shells and skeletons of microorganisms. Farther away from the ridge crest, the slope falls below the calcium carbonate compensation zone generally about 3 miles deep. Below this depth, calcium carbonate, whose solubility increases with pressure, dissolves in seawater. Therefore, only red clay should exist in the deep abyssal waters far from the crest of the midocean ridge. Yet drill cores taken from the abyssal plains near continental shelves, where the oceanic

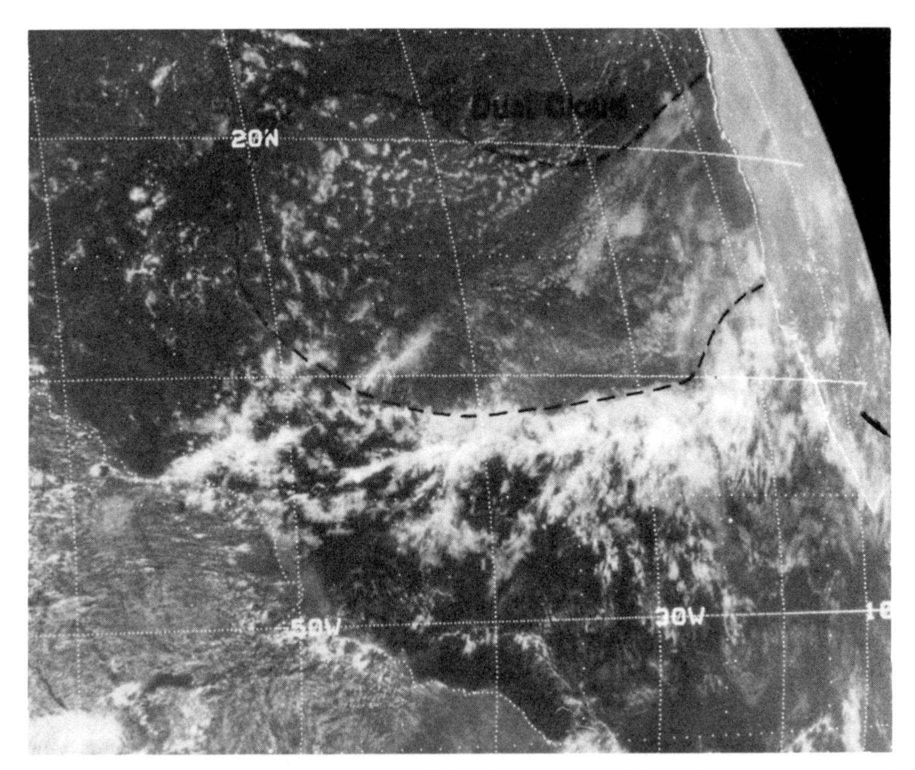

Figure 2–11 During the summer of 1976, drought conditions in West Africa and a prevailing easterly wind resulted in a dust surge—an enormous cloud of dust blowing out over the Atlantic Ocean (shown south of the dashed line) from the Sahara Desert. Courtesy of NOAA

crust is the oldest and deepest, clearly show thin layers of calcium carbonate below thick beds of red clay and above hard volcanic rock. Geologists concluded that the red clay protected the calcium carbonate from dissolving in the deep waters of the abyss. The discovery implies that the midocean ridge was the source of the calcium carbonate near continental margins and that the seafloor has moved across the Atlantic Basin.

MAGNETIC SURVEYS

Geologists looking for a decisive test for seafloor spreading stumbled upon magnetic reversals on the ocean floor. Experiments using sensitive magnetic recording instruments called magnetometers towed behind ships over the midocean ridges (Fig. 2–12) revealed magnetic patterns locked in the volcanic rocks on the seafloor. These patterns alternated from north to south and were mirror images of each other on opposite sides of the ridge crest. The magnetic fields captured in the rocks also showed the past position of the magnetic poles as well as their polarities.

Two or three times every million years, the Earth's geomagnetic field reverses polarity, with the north and south magnetic poles switching places. Over the last 4 million years, the filed has reversed 11 times. As the iron-rich basalts of the midocean ridges cool, the magnetic fields of their iron molecules line up in the direction of the Earth's magnetic field at the

Figure 2–12 A crewmember lowers a magnetometer over the stern of the oceanographic research ship USNS *Hayes*. Courtesy of U.S. Navy

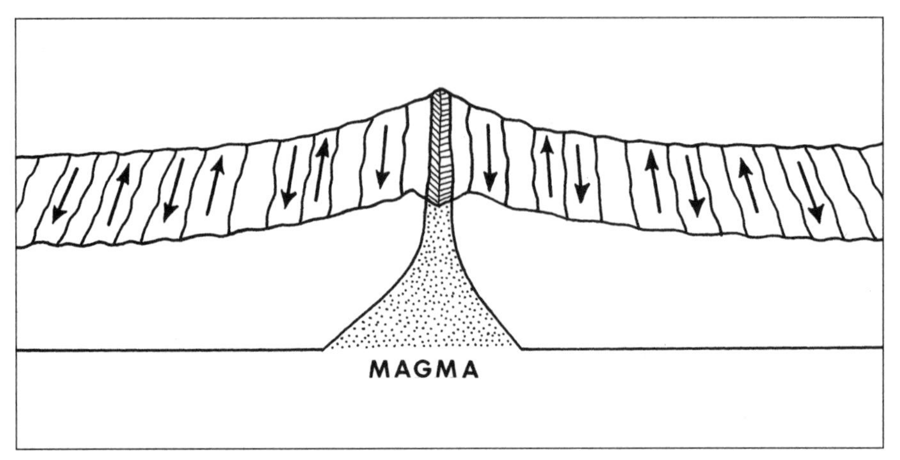

Figure 2–13 As volcanic rock cools at midocean ridges, it is polarized in the direction of the Earth's magnetic field, providing a series of magnetic stripes on the ocean floor.

time of their deposition. As the ocean floor spreads out on both sides of the ridge, the basalts solidify, establishing a record of the geomagnetic field at each successive reversal. This process produced parallel bands of magnetic rocks of varying width and magnitude on both sides of the ridge crest that were mirror images of each other (Fig. 2–13). Here at last was clinching proof for seafloor spreading: in order for the magnetic stripes to form in such a manner, the ocean floor had to be pulling apart.

TABLE 2–2 COMPARISON OF MAGNETIC REVERSALS WITH OTHER PHENOMENA (Dates in Millions of Years)

Magnetic Reversal	Unusual Cold	Meteorite Activity	Sea Level Drops	Mass Extinctions
0.7	0.7	0.7		
1.9	1.9	1.9		
2.0	2.0			
10				11
40			37–20	37
70			70–60	65
130			132–125	137
160			165–140	173

The magnetic stripes also provided a means of dating virtually the entire ocean floor, because the magnetic reversals occurred randomly, and any set of patterns is unique in geologic history. The rate of seafloor spreading was calculated by determining the age of the magnetic stripes by dating drill cores taken from the midocean ridge and measuring the distance from their points of origin at the ridge crest. During the past 100 million years, the rate of seafloor spreading has changed little. Periods of increased acceleration have been accompanied by an increase in volcanic activity. In the past 10 to 20 million years, there has been progressive accelleration, reaching a peak about 2 million years ago.

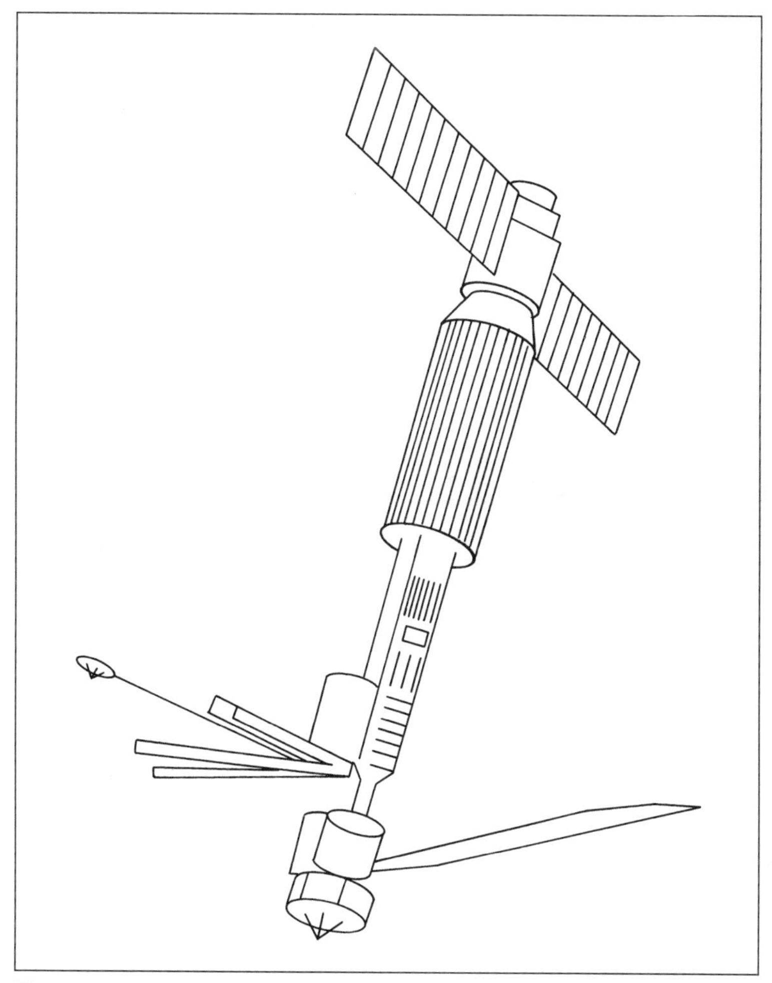

Figure 2–14 The Seasat satellite radar mapped the ocean surface over most of the globe.

The spreading rates on the East Pacific Rise are upward of 6 inches per year, which results in less topographical relief on the ocean floor. The active tectonic zone of a fast-spreading ridge is usually quite narrow, generally less than 4 miles wide. In the Atlantic, the rates are much slower, only about 1 inch per year, which allows taller ridges to form. Using the rate of seafloor spreading, the Atlantic appears to have opened around 170 million years ago—a time span remarkably concurrent with the estimated date for the breakup of the continents.

SATELLITE MAPPING

In 1978, the radar satellite *Seasat* (Fig. 2–14) precisely measured the distance to the ocean surface over most of the globe. Among the astonishing discoveries was the fact that ridges and trenches on the ocean bottom produce corresponding hills and valleys on the surface of the ocean because of variations in the pull of gravity. The topography of the ocean surface shows bulges and depressions, with several hundred feet of relief. But because these surface variations range over wide areas, they are unrecognized on the open sea.

The pull of gravity from undersea mountains, ridges, trenches, and other structures of varying mass distributed over the seafloor controls the shape of the surface water. Undersea mountain ranges produce large gravitational forces that cause seawater to pile up around them, resulting in gentle swells on the ocean surface. Conversely, submarine trenches with less mass to

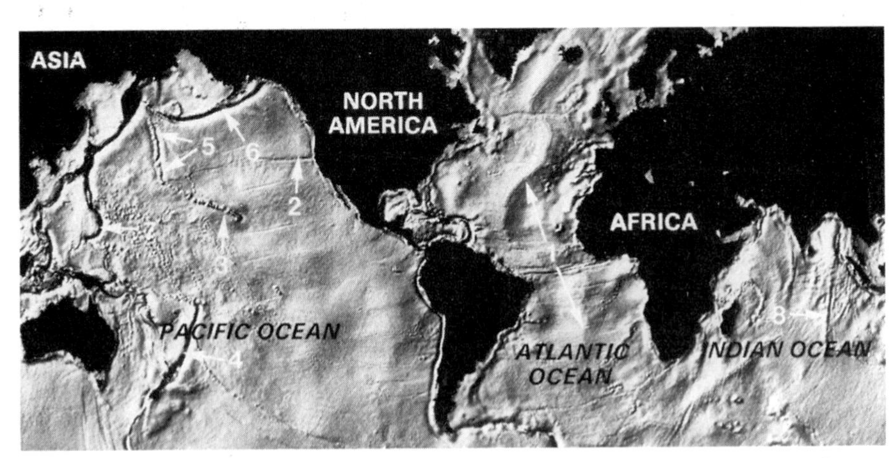

Figure 2–15 Radar altimeter data from the Geodynamic Experimental Ocean Satellite (GEOS-3) and *Seasat* was used to produce this map of the ocean floor. (1) Mid-Atlantic Ridge, (2) Mendocino Fracture Zone, (3) Hawaiian Island chains, (4) Tonga Trench, (5) Emperor Seamounts, (6) Aleutian Trench, (7) Mariana Trench, (8) Ninety East Ridge. Courtesy of NASA

attract water form shallow troughs in the sea surface. For example, a trench 1 mile deep can cause the ocean to drop dozens of feet.

The satellite altimetry data produced a map of the entire ocean surface (Fig. 2–15). Chains of midocean ridges and deep-sea trenches were delineated with a clarity greater than had been achieved by any other method of mapping the ocean floor. The seafloor maps also revealed many new features such as rifts, ridges, seamounts, and fracture zones and better defined several known features. These maps provided additional support for the theory of plate tectonics, which holds that the crust is broken into several plates whose constant shifting is responsible for the geologic activity on the Earth's surface, including the growth of mountain ranges and the widening of ocean basins.

The satellite imagery also revealed long-buried parallel fracture zones undiscovered by conventional seafloor mapping techniques. The faint lines running like a comb through the central Pacific seafloor might be influenced by convection currents in the mantle 30 to 90 miles beneath the oceanic crust. Each circulating loop consists of hot material rising and cooler material sinking back into the depths, tugging on the ocean floor as it descends.

Even buried structures came into full view for the first time. One example is an ancient midocean ridge that formed when South America, Africa, and Antarctica began separating around 125 million years ago. The seafloor spreading center was buried deep under thick layers of sediment. The boundary between the plates moved westward, leaving behind the ancient ridge, which began to subside. The ridge's discovery might help geologists trace the evolution of the oceans and continents over the last 200 million years. The satellite's discoveries are further proof that the deep-sea floor remains, in large part, uncharted territory, and that the exploration of inner space is as important as the exploration of outer space.

3

THE DYNAMIC SEAFLOOR

The ocean's crust is constantly changing. It is comparatively young, less than 5 percent of Earth's age. The age difference is due to the recycling of oceanic crust into the mantle; almost all the ocean floor has disappeared into the Earth's interior over the last 170 million years. The oceanic crust is continuously being created at midocean ridges, where basalt oozes out of the mantle through rifts in the crust, and destroyed in deep-sea trenches, where the lithosphere plunges into the mantle and remelts in a continuous cycle.

The divergence of lithospheric plates creates new oceanic crust at spreading ridges, while convergence destroys old oceanic crust in subduction zones. When two plates collide, the less buoyant oceanic crust subducts under continental crust. The lithosphere and the overriding oceanic crust recycle through the mantle to make new crust. The lithospheric plates act like rafts riding on a sea of molten rock, slowly carrying the continents around the surface of the globe.

LITHOSPHERIC PLATES

The Earth's outer shell is fractured like a broken egg into several large plates (Fig. 3–1). The shifting lithospheric plates range in size from a few hundred

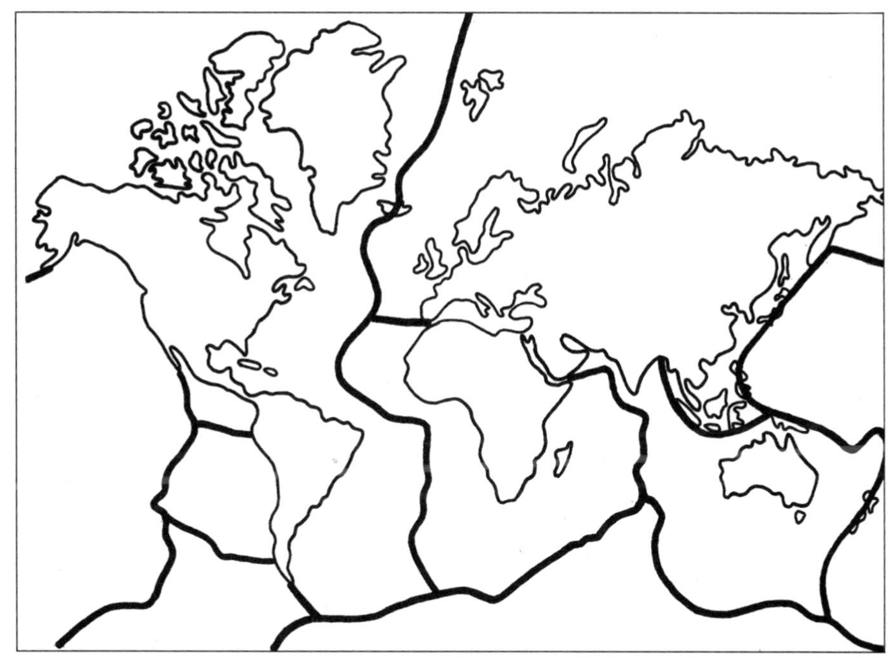

Figure 3–1 The major lithospheric plates.

to several million square miles. They comprise the crust and the upper brittle mantle called the lithosphere. The lithosphere consists of the rigid outer layer of the mantle and underlies the continental and oceanic crust. The thickness of the lithosphere is about 60 miles under the continents and averages about 25 miles under the ocean.

The lithospheric plates ride on a hot pliable layer of the upper mantle called the asthenosphere, in a manner similar to hard wax riding on melted wax. They carry the crust like drifting slabs of rock. The plates diverge at midocean spreading ridges and converge at subduction zones, which lie at the edges of lithospheric plates. The lithospheric plates subduct into the mantle in a continuous cycle of crustal regeneration. Their constant interaction with each other shapes the surface of the planet. This structure of the upper mantle is important for the operation of plate tectonics, which is responsible for all geologic activity on the Earth.

The plate boundaries are zones of active deformation that absorb the force of impact between nearly rigid plates. These boundary zones vary from a few hundred feet where plates slide past each other at transform faults to several tens of miles at midocean ridges and subduction zones. The divergent plate margins are midocean spreading ridges, where basalt welling up from within the upper mantle creates new oceanic crust as part of the process of seafloor spreading (Fig. 3–2).

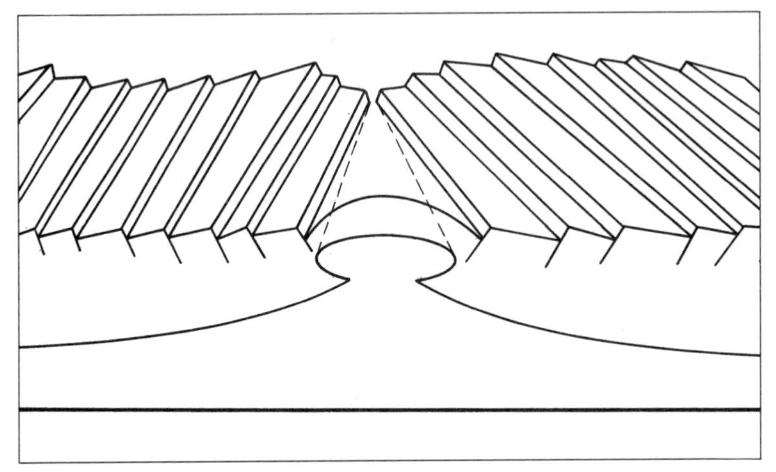

Figure 3–2 Creation of oceanic crust at a spreading ridge.

Oceanic crust does not form as a single homogeneous mass, but rather, is made in long narrow ribbons laid side by side and interspersed with fracture zones. The midocean ridge system, which does not always lie in the middle of the ocean, snakes 45,000 miles around the globe, making it the longest structure on Earth. The lateral plate margins are transform faults, where plates slide past each other accompanied by little or no tectonic activity such as the upwelling of magma and the generation of earthquakes.

The convergent plate margins are the subduction zones represented by deep-sea trenches, where old oceanic crust sinks into the mantle to provide magma for volcanoes fringing the trenches. If tied end to end, the subduction zones would stretch around the world. The convergence rates between plates range from less than 1 inch to more than 5 inches per year, corresponding to the rates of plate divergence. However, subduction zones and associated spreading ridges on the margins of a plate do not operate at the same rates; this disparity causes the plates to travel across the surface of the Earth. If subduction overcomes seafloor spreading, the lithospheric plate shrinks and eventually disappears altogether.

OCEANIC CRUST

The crust of the ocean is remarkable for its consistent thickness and temperature, averaging about 4 miles thick and not varying more than 20 degrees Celsius over most of the globe. By comparison, the continental crust is on average 25 to 30 miles thick, and in the domain of mountain ranges, it reaches a thickness of 45 miles. The continents also have thick roots of

TABLE 3–1 CLASSIFICATION OF THE EARTH'S CRUST

Environment	Crust Type	Tectonic Character	Thickness in Miles	Geologic Features
Continental crust overlying stable mantle	Shield	Very stable	22	Little or no sediment, exposed Pre-cambrian rocks
	Midcontinent	Stable	24	
	Basin & Range	Very unstable	20	Recent normal faulting, volcanism, and intrusion; high mean elevation
Continental crust overlying unstable mantle	Alpine	Very unstable	34	Rapid recent uplift, relatively recent intrusion; high mean elevation
	Island arc	Very unstable	20	High volcanism, intense folding and faulting
Oceanic crust overlying stable mantle	Ocean basin	Very stable	7	Very thin sediments overlying basalts, no thick Palaeozoic sediments
Oceanic crust overlying unstable mantle	Ocean ridge	Unstable	6	Active basaltic volcanism, little or no sediment

relatively cold mantle material extending down to a depth of about 250 miles. The average density of continental crust is 2.7 times the density of water, compared with 3.0 for oceanic crust and 3.4 for the mantle. The difference in density buoys up the continental and oceanic crust.

The oceanic crust is like a layer cake with 3 distinct strata. It has an upper layer of pillow basalts, formed when lava extruded undersea at great depths; a middle layer of a sheeted-dike complex, consisting of a tangled mass of feeders that bring magma to the surface; and a lower layer of gabbros, coarse-grained rocks that crystallized slowly under high pressure in deep magma chambers. The same rock formation is found on the continents. This similarity has led geologists to speculate that these formations were pieces of ancient oceanic crust called ophiolites (Fig. 3–3).

Most oceanic crust is less than 170 million years old, with a mean age of 100 million years, compared with the continental crust, which is about 4 billion years old. The difference in ages is due to the recycling of oceanic crust into the mantle, as described previously. Almost all the ocean floor has since disappeared into the Earth's interior to provide the raw materials for the continued growth of the continents.

As mentioned earlier, new oceanic crust forms at spreading ridges, where basalt oozes out of the mantle through rifts on the ocean floor. Some molten magma erupts as lava on the surface of the ridge through a system of vertical passages. Once at the surface, the liquid rock flows down the ridge and hardens into sheets or rounded forms of pillow lavas, depending on the rate of extrusion and the slope of the ridge. Periodically, lava overflows onto the ocean floor in gigantic eruptions, adding several square miles of new oceanic crust yearly. As the oceanic crust cools and hardens, it contracts, forming fractures through which water circulates.

Not all magma extrudes onto the ocean floor. Most of it cools and bonds to the edges of separating plates. Much of the magma solidifies within the conduits above the magma chambers, forming massive vertical sheets

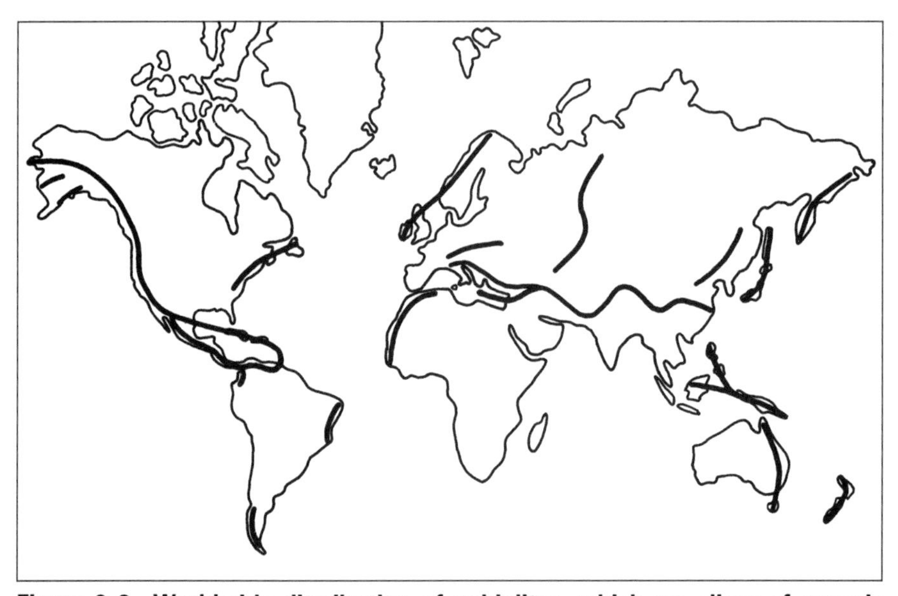

Figure 3–3 Worldwide distribution of ophiolites, which are slices of oceanic crust shoved up on land by the action of plate tectonics.

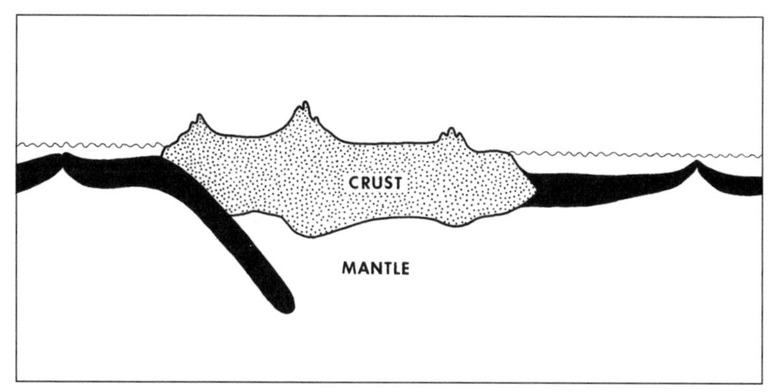

Figure 3–4 The Earth's crust is composed of continental granites and ocean basin basalts.

called dikes that resemble a deck of cards standing on end. Individual dikes measure about 10 feet thick, about 1 mile high, and approximately 3 miles long.

The oceanic plates thicken with age, from a few miles thick after formation at midocean spreading ridges to more than 50 miles thick in the oldest ocean basins next to the continents. The depth at which an oceanic plate sinks as it moves away from a midocean spreading ridge varies at a ratio of the square root of its age. For example, a plate that is 2 million years old lies about 2 miles deep; a plate that is 20 million years old lies about 2.5 miles deep; and a plate that is 50 million years old lies about 3 miles deep.

A typical oceanic plate starts out thin and gradually thickens by the underplating of new lithosphere from the upper mantle and the accumulation of overlying sediment layers. The ocean floor at the summit of a midocean ridge consists almost entirely of hard basalt and acquires a thickening layer of sediments farther outward from the ridge crest. By the time the oceanic plate extends as wide as the Atlantic Ocean, the portion near continental margins where the sea is the deepest is about 60 miles thick. Eventually, the oceanic plate becomes so thick and heavy that it can no longer remain on the surface. It then bends downward and subducts beneath a continent into the Earth's interior (Fig. 3–4).

As the oceanic plate dives into a subduction zone, it remelts and acquires new minerals from the mantle, providing the raw material for new oceanic crust in the form of molten magma that reemerges at volcanic spreading centers along midocean ridges. Sediments deposited on the ocean floor and the water trapped between sediment grains are also caught in the subduction zones. But the lower melting points and lesser density of these molten sediments makes them rise toward the surface to supply nearby volcanoes with magma and recycled seawater.

The fluid portion of the upper mantle is called the asthenosphere. Here rocks are semimolten or plastic, enabling them to slowly flow. After millions of years, the molten rocks reach the topmost layer of the mantle, or lithosphere. With a reduction of pressure within the Earth, the rocks melt and rise through fractures in the lithosphere. As the molten magma passes through the lithosphere, it reaches the bottom of the oceanic crust, where it forms magma chambers that further press against the crust, which continues to widen the rift. Molten lava pouring out of the rift forms ridge crests on both sides and adds new material to the spreading ridge system (Fig. 3–5).

The mantle material below spreading ridges where new oceanic crust forms is mostly peridotite, a strong, dense rock composed of iron and magnesium silicates. As the peridotite melts on its journey to the base of the oceanic crust, a portion becomes highly fluid basalt, which is the most common magma erupted on the surface of the Earth. About 5 cubic miles of basaltic magma is removed from the mantle and added to the crust every year. Most of this volcanism occurs on the ocean floor at spreading centers, where the oceanic crust pulls apart. Gabbro containing higher amounts of silica solidifies out of the basaltic melt and accumulates in the lower layer of the oceanic crust.

The oceanic crust, composed of basalts originating at spreading ridges and sediments washed off continents and islands, gradually increases density and finally subducts into the mantle. On its way deep into the Earth's interior, the lithosphere and its overlying sediments melt. The molten magma rises toward the surface in huge bubblelike structures called diapirs, from the Greek word meaning "to pierce." When the magma

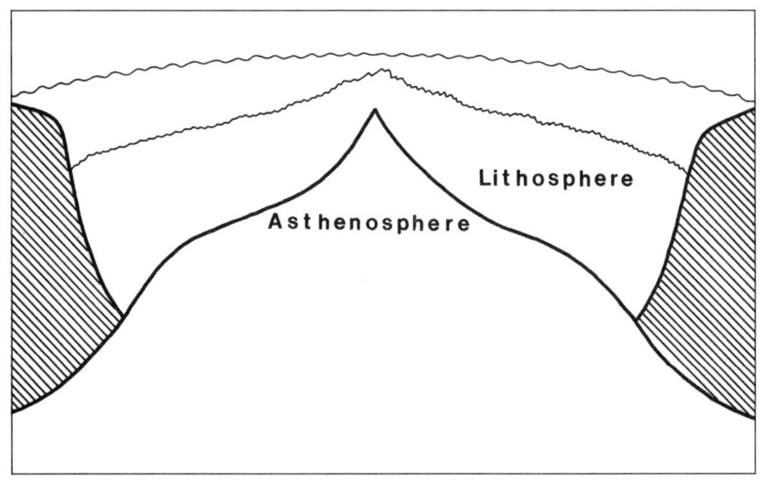

Figure 3–5 The structure of a spreading ridge, where material from the asthenosphere produces new lithosphere.

reaches the base of the crust, it provides new molten rock for magma chambers beneath volcanoes and granitic bodies called plutons, which often form mountains. In this manner, plate tectonics is continually changing and rearranging the face of the Earth.

THE ROCK CYCLE

The entire volume of the world's oceans circulates through the crust at spreading ridges every 10 million years, a volume approximately equivalent to the annual flow of the Amazon, the world's largest river. This action accounts both for the unique chemistry of seawater and for the efficient thermal and chemical exchanges between the crust and the ocean. The magnitude of some of these chemical exchanges is comparable in volume to the input of elements into the oceans by all the world's rivers, which carry materials weathered from the continents. The most important of these chemical elements is carbon, which controls many of the life processes on the planet.

When the seafloor subducts into the Earth's interior, the intense heat of the mantle drives out carbon dioxide from carbonaceous sediments. The molten rock, with its contingent of carbon dioxide, works its way upward through the mantle and fills the magma chambers that underly volcanoes and spreading ridges. The consequent eruption of volcanoes and the flow of molten rock from midocean ridges resupplies the atmosphere and ocean

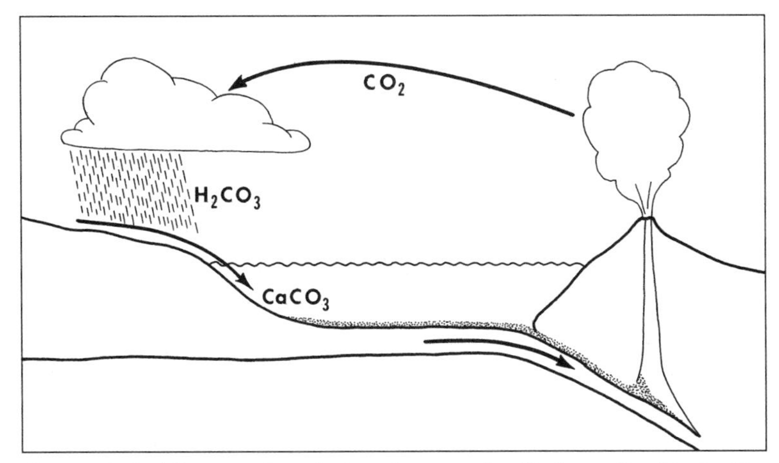

Figure 3-6 The geochemical carbon cycle. Carbon dioxide in the form of bicarbonate is washed off the land and enters the ocean, where organisms convert it to carbonate sediments, which are thrust into the mantle, become part of the magma, and escape into the atmosphere from volcanoes.

TABLE 3–2 RELATIVE AMOUNTS OF CARBON

Source	Relative Amount
Calcium carbonate in sedimentary rocks	60,000
Ca-Mg carbonate in sedimentary rocks	45,000
Sedimentary organic matter in the remains of animal tissues	25,000
Bicarbonate and carbonate dissolved in ocean	75
Coal and petroleum	7
Soil humus	5
Atmospheric carbon dioxide	1.5
All living plants and animals	1

with new carbon dioxide, making the Earth in effect one great carbon dioxide recycling plant (Fig. 3–6).

The geochemical carbon cycle—the transfer of carbon within the Earth—involves the interactions between the crust, ocean, atmosphere, and life. The biological carbon cycle is only a small component of this cycle, and is the transfer of carbon from the atmosphere to vegetation by photo-

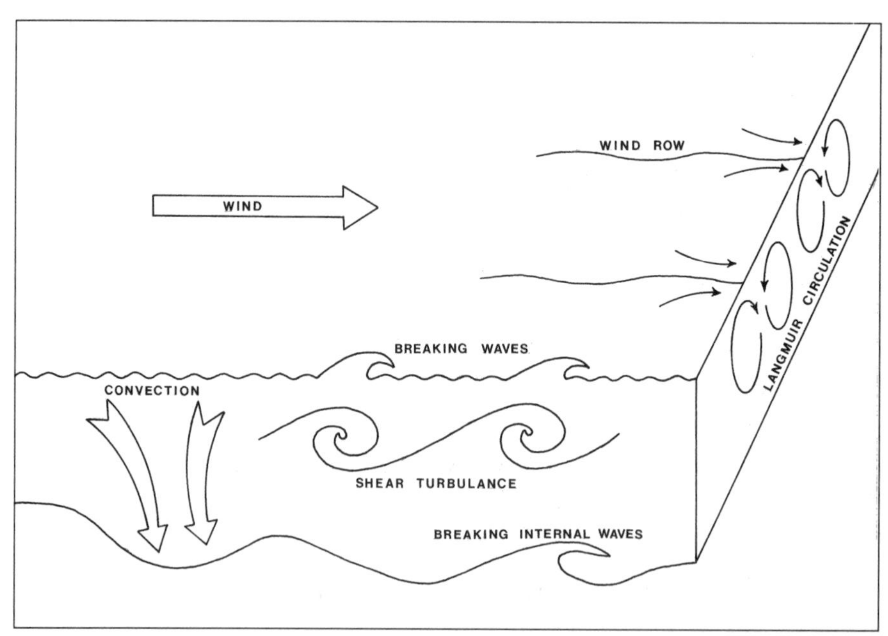

Figure 3–7 Turbulence in the upper layers of the ocean induces the mixing of gases.

synthesis, returning carbon to the atmosphere when plants respire or decay. The vast majority of carbon is not stored in organic matter, however, but is locked up in sedimentary rocks on the ocean floor and on the continents.

The oceans play a critical role in the carbon cycle by regulating the level of carbon dioxide in the atmosphere. In the upper layers of the ocean, the concentration of gases is in equilibrium with the atmosphere: the mixed layer of the ocean within the upper 300 feet (Fig. 3–7) contains as much carbon dioxide as the entire atmosphere. The gas dissolves into seawater mainly by the agitation of surface waves. Without marine photosynthetic organisms to absorb dissolved carbon dioxide, much of the reservoir of this gas would escape into the atmosphere, more than tripling the present carbon dioxide content and causing a runaway greenhouse effect.

Atmospheric carbon dioxide combines with rain to form carbonic acid. The acid reacts with surface rocks, producing dissolved calcium and bicarbonate, which are carried by streams to the ocean. Marine organisms use these substances to build calcium carbonate skeletons and other supporting structures. When the organisms die, their skeletons sink to the bottom of the ocean, where they dissolve in the deep abyssal waters. Because of its large volume, the abyss holds the largest reservoir of carbon dioxide in the world.

In shallow water, the carbonate skeletons build deposits of limestone (Fig. 3–8), which buries carbon dioxide in the geologic column comprising

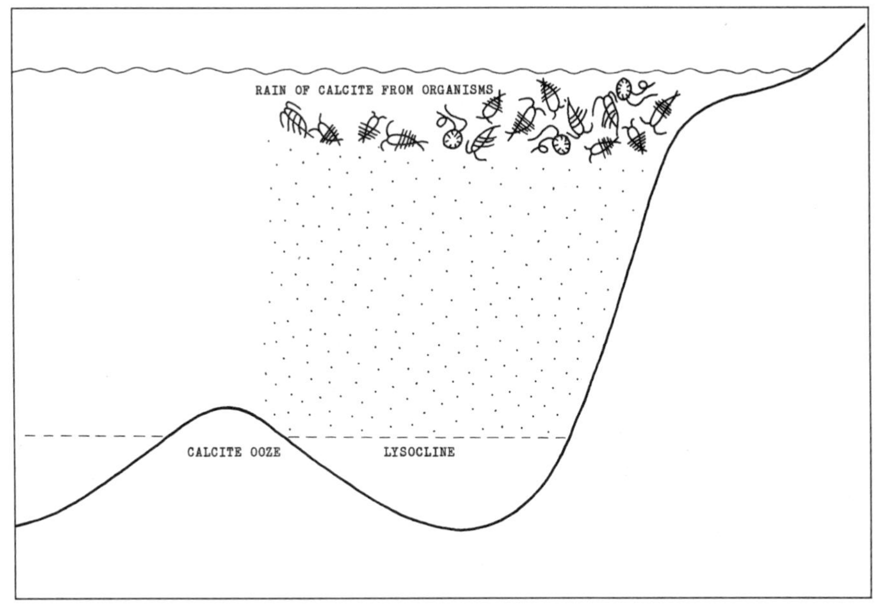

Figure 3–8 Formation of carbonate sediments on the ocean floor from the burial of shells and skeletons of marine organisms.

all sedimentary rocks. The burial of carbonate is responsible for about 80 percent of the carbon deposited on the ocean floor. The remainder of the carbonate originates from dead organic matter washed off the continents. Half the carbonate transforms back into carbon dioxide, which escapes into the atmosphere. Without the return of carbon dioxide, photosynthesis would cease and all life would end.

The deep water, which represents about 90 percent of the ocean's volume, circulates very slowly, with a residence time (time in place) of about 1,000 years. It comes into direct contact with the atmosphere only in the polar regions. Thus, the deep water's absorption of carbon dioxide is limited. The abyss receives most of its carbon in the form of the shells of dead organisms and fecal matter that sink to the ocean bottom. Carbon dioxide returns to the atmosphere by upwelling currents in the tropics, which is why the concentration of this gas is greater near the equator.

Volcanic activity on the ocean floor and on the continents plays a vital role in restoring the carbon dioxide content of the atmosphere. Carbon dioxide escapes from carbonaceous sediments that melt in the Earth's interior to provide new magma. The molten magma, along with volatiles including water and carbon dioxide, rises to the surface to feed magma chambers beneath midocean ridges and volcanoes. When the volcanoes erupt, carbon dioxide is released from the magma and returns to the atmosphere, completing the cycle.

OCEAN BASINS

Ocean basins are the largest depressions on Earth. The ocean floor lies much deeper below sea level than the continents rise above it. If the oceans were completely drained of water, the planet would look much like the rugged surface of Venus, which lost its oceans eons ago. The deepest parts of the dry seabed would lie several miles below the surrounding continental margins. The floor of the desiccated ocean would be traversed by the longest mountain ranges and fringed in many places by the deepest trenches. Vast empty basins would divide the continents, which would stand out like thick slabs of rock.

Most of the seawater that surrounds the continents lies in a single great basin in the Southern Hemisphere, which is nine-tenths ocean. It branches northward into the Atlantic, Pacific, and Indian basins in the Northern Hemisphere, where most of the continental landmass exists. The Arctic Ocean is a nearly landlocked sea connected to the Atlantic and Pacific only by narrow straits. The Bering Sea (Fig. 3–9) separates Alaska and Asia by only 56 miles at its narrowest point. About 20 million years ago, a ridge near Iceland subsided, allowing cold water from the recently formed Arctic Ocean to surge into the Atlantic, giving rise to the oceanic circulation system in existence today.

TABLE 3–3 HISTORY OF DEEP CIRCULATION IN THE OCEAN

Age (million years ago)	Event
3	An ice age overwhelms the Northern Hemisphere.
3–5	Arctic glaciation begins.
15	The Drake Passage is fully opened; the circum-Antarctic current is formed. Major sea ice forms around Antarctica, which is glaciated, making it the first major glaciation of the Modern Ice Age. The Antarctic bottom water forms. The snow limit rises.
25	The Drake Passage between South America and Antarctica begins to open.
25–35	A stable situation exists with possible partial circulation around Antarctica. The equatorial circulation is interrupted between the Mediterranean Sea and the Far East.
35–40	The equatorial seaway begins to close. There is a sharp cooling of the surface and of the deep water in the south. The Antarctic glaciers reach the sea with glacial debris in the sea. The seaway between Australia and Antarctica opens. Cooler bottom water flows north and flushes the ocean. The snow limit drops sharply.
>50	The ocean flows freely around the world at the equator. Rather uniform climate and warm ocean exists even near the poles. Deep water in the ocean is much warmer than it is today. Antarctica only has alpine glaciers, with no major ice sheets.

The oceans expand across some 70 percent of the Earth's surface, covering an area of about 140 million square miles with more than 300 million cubic miles of seawater. About 60 percent of the planet is covered by water no less than 1 mile deep, with an average depth of about 2.3 miles. The midocean ridges lie at an average depth of 1.5 miles, and the ocean bottom slopes away on both sides to a depth of about 3.5 miles. In the Pacific Basin, the ocean is up to 7 miles deep, the lowest point on Earth.

With only marine-born sedimentation and no bottom currents to stir up the seabed, an even blanket of material would settle onto the original volcanic floor of the oceans. Instead, however, the rivers of the world contribute a substantial amount of the sediment deposited on the deep ocean floor. The largest rivers of North and South America empty into the Atlantic, which receives considerably more river-borne sediment than does

Figure 3–9 St. Lawrence Island in the Bering Sea, showing cinder cones at the northwest end of the Kookooligit Mountains. Photo by H. B. Allen, courtesy of USGS

the Pacific. The burial of organic material also greatly aids the formation of offshore petroleum reserves.

Because the Atlantic is smaller and shallower than the Pacific, its marine sediments are buried more rapidly and therefore are more likely to survive than those in the Pacific. The floor of Atlantic accumulates sediments at a rate of about an inch every 2,500 years. The deep-ocean trenches around the Pacific trap much of the material reaching its western edge, where it subducts into the mantle.

Strong near-bottom currents redistribute sediments in the Atlantic on a greater scale than in the Pacific. Abyssal storms with powerful currents occasionally sweep patches of ocean floor clean of sediments and deposit the debris elsewhere. On the western side of the Pacific, Atlantic, and Indian ocean basins, periodic undersea storms skirt the foot of the continental rise and transport huge loads of sediment, dramatically modifying the seafloor. The scouring of the seabed and deposition of thick layers of fine sediment results in a much more complex marine geology than would be developed simply by a constant rain of sediments from above.

SUBMARINE CANYONS

The ocean floor presents a rugged landscape unmatched anywhere else on Earth. Chasms dwarfing even the largest continental canyons plunge to great depths. Rivers emptying into the sea eroded the exposed seabed when the sea level lowered dramatically during the last ice age.

At the height of the last ice age, about 10 million cubic miles of the Earth's water were held in the continental ice sheets, which covered about a third of the land surface with an ice volume three times greater than its present size. The accumulated ice dropped the level of the ocean by about 400 feet, advancing the shoreline hundreds of miles seaward. The coastline of the eastern seaboard of the United States extended about halfway to the edge of the continental shelf, which runs eastward more than 600 miles. The drop in sea level exposed land bridges and linked continents.

Numerous canyons slice through the continental shelf beneath the Bering Sea between Alaska and Siberia. About 75 million years ago, continental movements created the broad Bering shelf rising 8,500 feet above the deep ocean floor. The shelf was exposed as dry land at several times during the ice ages when sea levels dropped hundreds of feet, and terrestrial canyons cut deep into the shelf. When the ocean refilled again at the end of the last

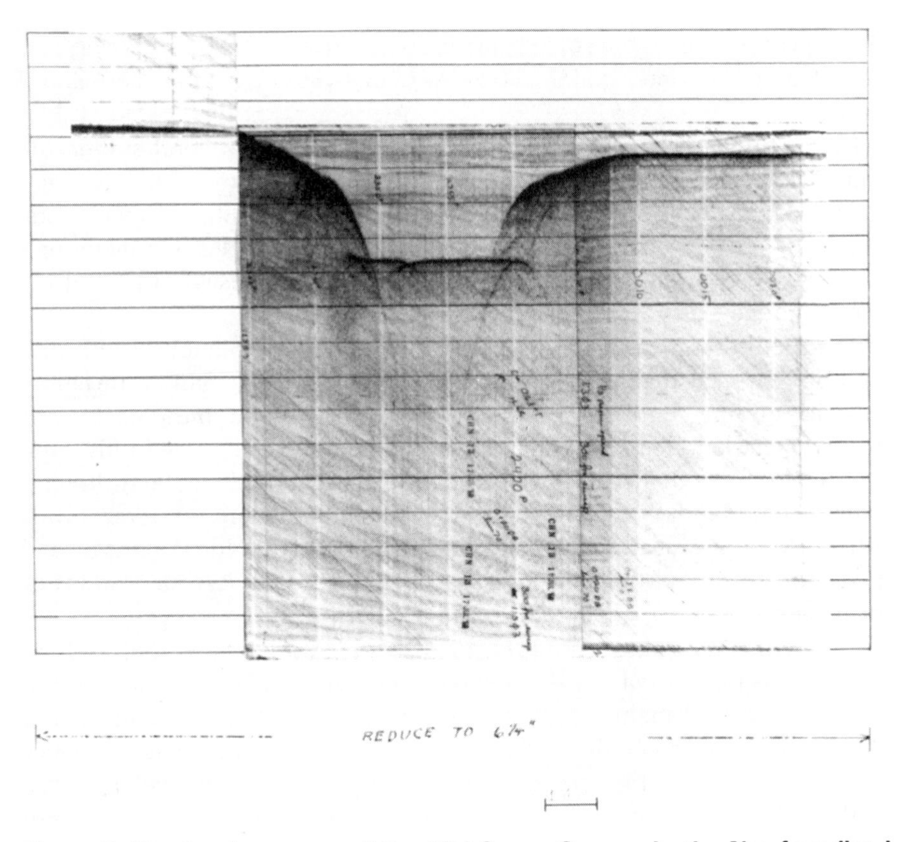

Figure 3–10 A seismogram of the Mid-Ocean Canyon in the Newfoundland Basin. Photo by R. M. Pratt, courtesy of USGS

ice age, massive landslides and mudflows swept down steep slopes on the shelf's edge, gouging out 1,400 cubic miles of sediment and rock.

A step resembling a sea cliff on the continental shelf off the eastern United States has been traced for nearly 200 miles. It appears to represent the former ice age coastline, now completely submerged under seawater. The massive continental glaciers that sprawled over much of the Northern Hemisphere held enough water to lower the sea by several hundred feet. When the glaciers melted, the sea returned to near its present level. Submarine canyons carved into bedrock 200 feet below sea level can be traced to rivers on land.

Several submarine canyons slice through the continental margin and ocean floor off eastern North America (Fig. 3–10). Submarine canyons on continental shelves and slopes possess many features identical to those of river canyons, and some rival even the largest on the continents. These canyons are characterized by high, steep walls and an irregular floor that slopes continually outward. The canyons are upward of 30 miles or more in length, with an average wall height of about 3,000 feet. The Great Bahamas Canyon is one of the largest submarine canyons, with a wall height of 14,000 feet, making it more than twice as deep as the Grand Canyon.

Rivers flowing across the exposed land gouged out several submarine canyons in the ocean floor when sea levels were much lower than they are today. Many submarine canyons have heads near the mouths of large rivers. Some submarine canyons extend to depths of over 2 miles, too deep for a terrestrial river origin. They formed instead by undersea slides, which carve out deep gashes in the ocean floor.

The Mediterranean Sea appears to have almost completely dried up 6 million years ago, making its seafloor a desert basin more than a mile below the surrounding continental plateaus. Rivers draining into the desiccated basin gouged out deep canyons. A deep sediment-filled gorge follows the course of the Rhône River in southern France for more than 100 miles and extends to a depth of 3,000 feet below the surface where the river drains into the Mediterranean Sea. Under the sediments of the Nile Delta is buried a mile-deep canyon that can be traced as far south as Aswan, 750 miles away, and is comparable in size to the Grand Canyon.

Submarine slides move rapidly down steep continental slopes and are responsible for excavating deep submarine canyons. The surface of the slopes is covered mainly with fine sediments swept off the continental shelves by submarine slides. The slides consist of sediment-laden water that is denser than the surrounding seawater. The turbid water moves swiftly along the ocean floor, eroding the soft bottom material. These muddy waters, called turbidity currents, move down steep slopes and play a major role in shifting the sands of the deep sea (Fig. 3–11).

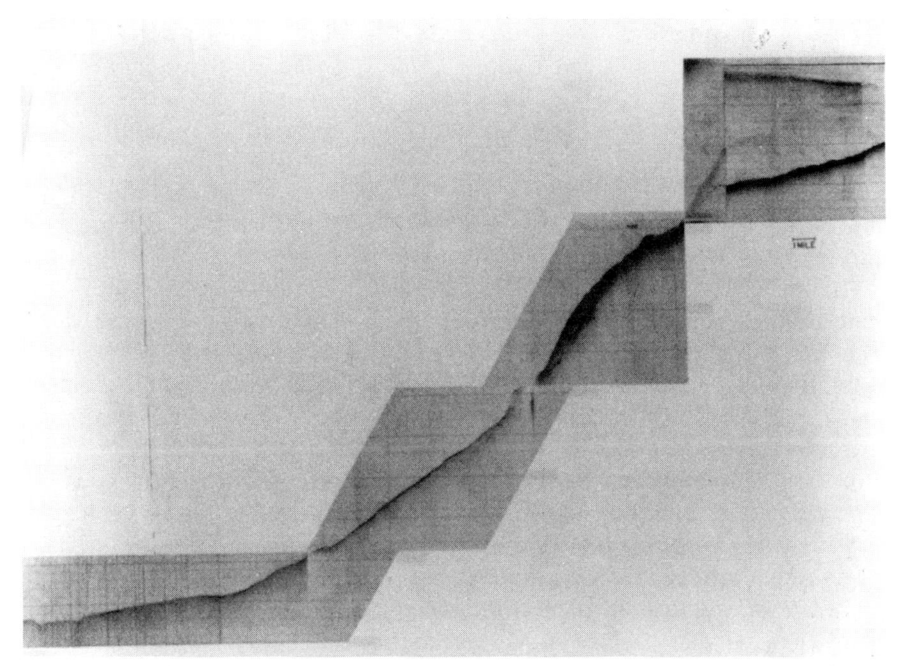

Figure 3–11 Sonar profile of the continental slope south of Nantucket, showing slumped debris from 5,000- to 7,500-foot depths. Photo by R. M. Pratt, courtesy of USGS

MICROPLATES AND TERRANES

During the initial breakup of Pangaea in the early Jurassic period, the Pacific plate—the largest in the world—was hardly bigger than the present-day continental United States. About 190 million years ago, the Pacific plate might have begun as a tiny microplate, a small block of oceanic crust that sometimes lies at the junction between two or three major plates. The rest of the ocean floor consisted of other unknown plates that have long since disappeared as the Pacific plate continued to grow. This is why no oceanic crust is older than Jurassic in age.

A microplate about the size of Ohio sits at the junction of the Pacific, Nazca, and Antarctic plates in the South Pacific about 2,000 miles west of South America. Seafloor spreading along the boundary zone between the plates adds new oceanic crust onto their edges, causing the plates to diverge. The different rates of seafloor spreading have caused the microplate at the hub of the spreading ridges, which fan out like the spokes of a bicycle wheel, to rotate one quarter-turn clockwise in the last 4 million years. A similar microplate near Easter Island to the north has spun nearly 90 degrees over the last 3 to 4 million years, suggesting that most microplates behave in this manner.

Three lithospheric plates bordering the Pacific Ocean—the Nazca, Antarctic, and South American plates—come together in an unusual triple junction. The first two plates spread apart along a boundary called the Chile Ridge, off the west coast of South America, similar to the way the Americas drift away from Eurasia and Africa along the Mid-Atlantic Ridge. The Chile Ridge lies off the Chilean continental shelf at a depth of more than 10,000 feet. Along its axis, magma rises from deep within the Earth and piles up into mounds forming undersea volcanoes.

The Nazca plate moving northeast subducts beneath the westward-moving South American plate at the Peru-Chile Trench. The eastern edge of the Nazca plate is subducting at a rate of about 50 miles every million years, which is faster than its western edge is growing. In essence, the Chile Trench is consuming the Chile Ridge, which will eventually disappear altogether. Several times in the past 170 million years, other plates and their associated spreading centers have vanished beneath the continents that surround the Pacific Basin. This activity had a substantial impact on the coastal geology.

A high degree of geologic activity around the rim of the Pacific Basin formed virtually all the mountain ranges facing the Pacific, as well as the island arcs along its perimeter. Much of western North America assembled from island arcs and other crustal debris skimmed off the Pacific plate as the North American plate moved westward. Northern California is a jumble of crustal fragments assembled over 100 million years ago. Rock formations in San Francisco came from as far as 2,500 miles across the Pacific Ocean.

Figure 3–12 Maclaren Glacier on the south side of the Alaskan Range. Photo by T. L. Pewe, courtesy of USGS

A nearly complete slice of ocean crust, the type that shoves up on the continents by drifting plates, sits in the middle of Wyoming. The entire state of Alaska is an assemblage of about 50 terranes set adrift over the past 160 million years by the wanderings and collisions of crustal plates (Fig. 3–12). Similarly, the Andes might have thrust upward by the accretion of oceanic plates along the continental margin of South America.

Terranes are patches of oceanic crust originating from faraway sources shoved up onto the continents and assembled into geologic collages. They are distinct from their geologic surroundings and are usually bounded by faults. The composition of terranes generally resembles that of an oceanic island or plateau, although some comprise a consolidated conglomerate of pebbles, sand, and silt that accumulated in an ocean basin between colliding crustal fragments.

Most terranes created on an oceanic plate are elongated bodies that deformed when colliding with a continent, which subjects them to crustal

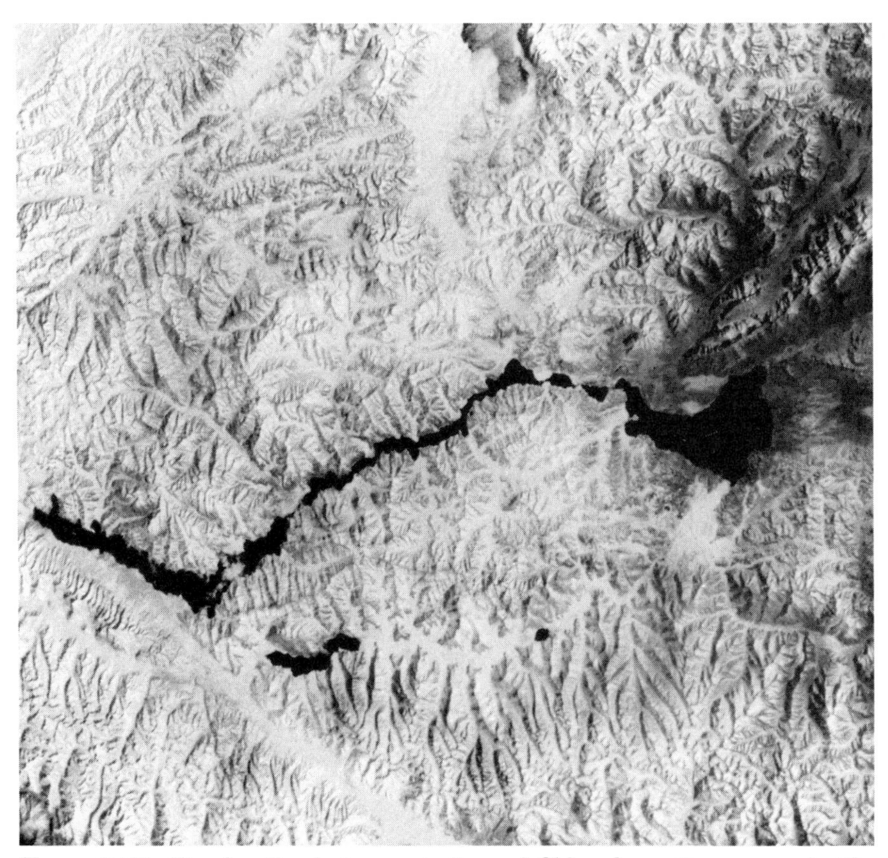

Figure 3–13 The frontier between India and China from the space shuttle, showing the Himalaya Mountains. Courtesy of NASA

movements that modify their shape. They exist in a variety of shapes and sizes, from small slivers to subcontinents such as India. The assemblage of terranes in China is being stretched and displaced in an east-west direction because of the continuing pressure that India exerts on southern Asia as it raises the Himalayas (Fig. 3–13). North of the Himalayas lies a belt of ophiolites, which appears to mark the boundary between the sutured continents. Ophiolites (from the Greek word *ophis*, meaning "serpent") are slices of ocean floor shoved up on the continents by drifting plates and date as old as 3.6 billion years.

Suspect terranes are fault-bounded blocks whose geologic histories are distinct from those of neighboring terranes and of adjoining continental masses. They range in age from less than 200 million years old to well over a billion years old. Different species of fossil radiolarians—marine protozoans with skeletons of silica and abundant from about 500 million to 160 million years ago—determine the age of the terranes and also defined specific regions of the ocean where the terranes originated.

Many terranes that comprise western North America have rotated clockwise as much as 70 degrees or more, with the oldest terranes having the greatest degree of rotation. Terrane boundaries, called suture zones, are commonly marked by ophiolite belts, consisting of marine sedimentary rocks, pillow basalts, sheeted dike complexes, gabbros, and peridotites. Suspect terranes were displaced over great distances before finally coming to rest at a continental margin. Some North American suspect terranes have a western Pacific origin and were displaced thousands of miles to the east. At their usual rate of travel, terranes could make a complete circuit of the globe in only half a billion years.

4

RIDGES AND TRENCHES

Vast undersea mountain ranges, much more extensive than those on the continents, crisscross the ocean stretches. A continuous system of midocean ridges girdles the planet, and is by far the longest geologic structure. Although deeply submerged, the midocean ridge system is easily the Earth's most dominant feature, extending over a larger area than all major terrestrial mountain ranges combined.

The subduction of the lithosphere in deep-sea trenches plays a fundamental role in global tectonics and accounts for powerful geologic forces that continuously shape the surface of the Earth. Major mountain ranges and most volcanoes are associated with the subduction of lithospheric plates. The subduction of the oceanic crust into the mantle produces strain in the descending lithosphere, causing powerful earthquakes to rumble across the landscape.

THE MIDOCEAN RIDGES

The shifting lithospheric plates create new oceanic crust in a continuous cycle of crustal rejuvenation. The subducting lithosphere circulates through the mantle and reemerges as magma at a dozen or so midocean ridges around the world, generating more than half the Earth's crust. The addition of new basalt to the ocean floor is responsible for the growth of the lithospheric plates upon which the continents ride.

A large part of this activity takes place in the middle of the Atlantic Ocean, where molten rock welling up from the upper mantle generates new sections of oceanic crust. The floor of the Atlantic acts like two conveyor belts, whose rollers are convection loops in the upper mantle, transporting oceanic crust in opposite directions outward from its point of origin at the Mid-Atlantic Ridge (Fig. 4–1).

The spreading ridge system runs from Iceland in the north to Bovet Island (about 1,000 miles off Antarctica) in the south. The midocean ridge is a string of volcanic seamounts, created by molten magma upwelling from within the mantle. Running down the middle of the ridge crest is a deep trough like a giant crack in the ocean's crust. This trough reaches 4 miles deep and is up to 15 miles wide, making it the greatest chasm on Earth.

The submerged mountains and undersea ridges form a continuous chain 45,000 miles long (Fig. 4–2). The mountainous belt is several hundred miles wide and rises upward of 10,000 feet above the ocean floor. Starting out from the Arctic Ocean, the ridge system spans southward across the

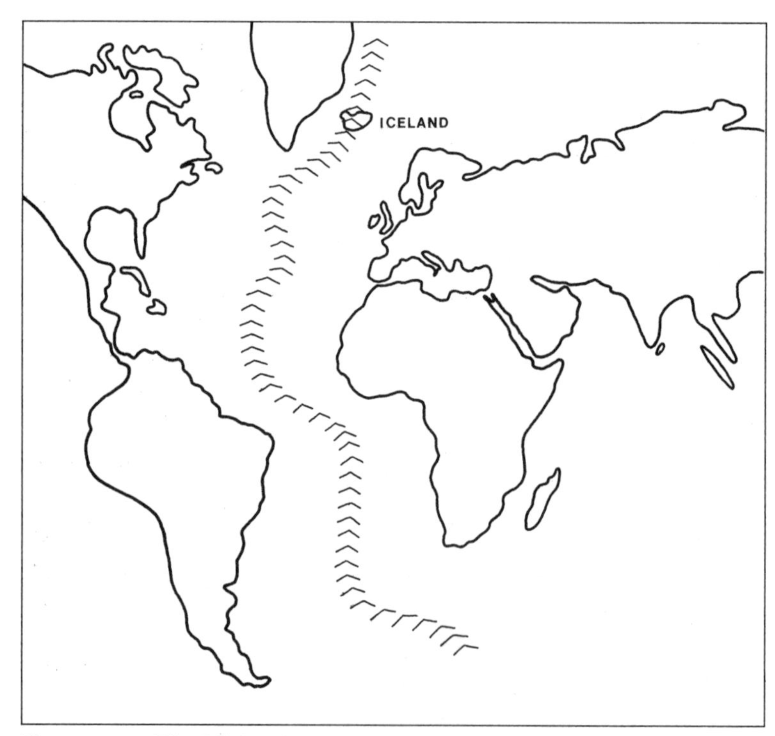

Figure 4–1 The Mid-Atlantic spreading ridge system separated the New World from the Old World.

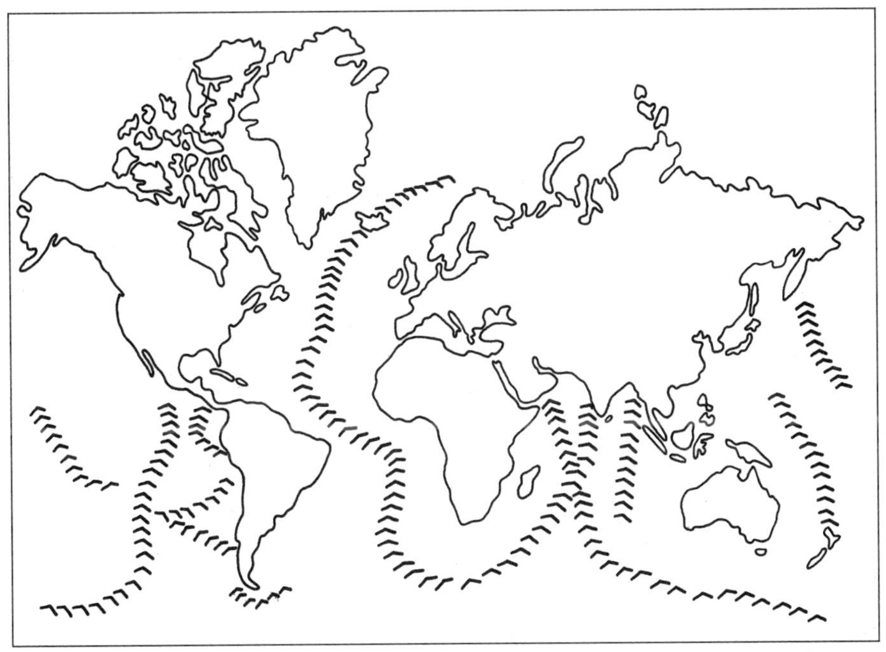

Figure 4–2 Midocean ridges that wind around the world's ocean basins are composed of individual volcanic spreading centers.

Atlantic Basin, continues around Africa, Asia, and Australia, runs under the Pacific Ocean, and terminates at the west coast of North America.

The ocean floor at the crest of the ridge consists mainly of basalt, the most common magma erupted on the surface of the Earth. About 5 cubic miles of new basalt is added to the crust annually, mostly on the ocean floor at spreading ridges. With increasing distance from the crest, a thickening layer of sediments shrouds the bare volcanic rock. As the two newly separated plates move away from the rift, material from the asthenosphere adheres to their edges to form new lithosphere. The lithospheric plate thickens as it propagates from a midocean rift system, causing the plate to sink deeper into the mantle; this is why the seafloor near the continental margins surrounding the Atlantic Basin is the deepest part of the ocean.

Intense seismic and volcanic activity along the midocean ridges manifests itself as a high heat flow from the Earth's interior. Molten magma originating from the mantle rises through the lithosphere and adds new basalt to both sides of the ridge crest. The greater the flow of magma, the more rapid the seafloor spreading and the lower the relief. The spreading ridges in the Pacific Ocean are more active than those in the Atlantic and therefore are less elevated. Rapid spreading ridges do not achieve the heights of slower ones because the magma does not have the opportunity to pile up into tall heaps. The axis of a slow-spreading ridge is characterized by a rift valley several miles deep and about 10 to 20 miles wide.

In the Pacific Ocean, a rift system called the East Pacific Rise stretches 6,000 miles from the Antarctic Circle to the Gulf of California. It lies on the eastern edge of the Pacific plate, marking the boundary between the Pacific and Cocos plates. It is the counterpart of the Mid-Atlantic Ridge and a member of the world's largest undersea mountain chain. The East Pacific Rise is a network of midocean ridges, which lie mostly at a depth of about 1.5 miles. Each rift is a narrow fracture zone, where plates of the oceanic crust diverge at an average rate of about 5 inches a year.

A set of closely spaced fracture zones dissects the Mid-Atlantic Ridge in the equatorial Atlantic. The largest of these structures is the Romanche Fracture Zone, which offsets the axis of the ridge in an east-west direction by nearly 600 miles (Fig. 4–3). The floor of the Romanche trench is as much as 5 miles below sea level, and the highest parts of the ridges on either side of the trench are less than a mile below sea level, providing a vertical relief four times that of the Grand Canyon.

The shallowest portion of the ridge is capped with a fossil coral reef, suggesting it was above sea level some 5 million years ago. Many similar and equally impressive fracture zones span the area, culminating in a sequence of troughs and transverse ridges several hundred miles wide. The resulting terrain is unmatched in size and ruggedness anywhere else in the world.

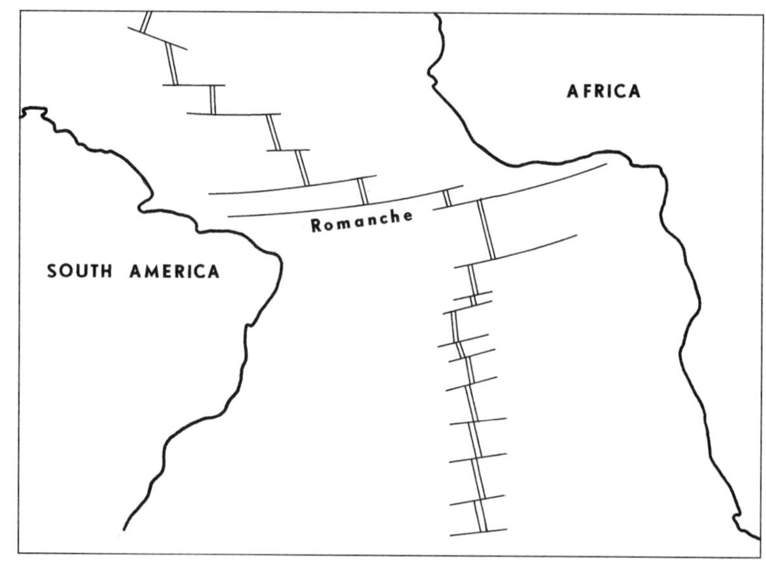

Figure 4–3 The Romanche Fracture Zone, which offsets the Mid-Atlantic Ridge.

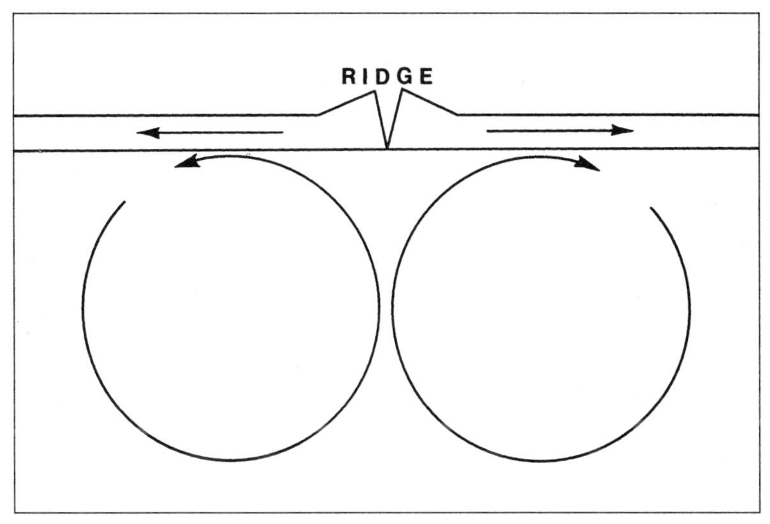

Figure 4–4 Convection currents in the mantle spread lithospheric plates apart.

THE HEAT ENGINE

All geologic activity that shapes the surface of the Earth is an outward expression of the great heat engine in the interior of the planet. The motion of the mantle churning over ever so slowly below the crust brings heat from the core to the surface in convection loops (Fig. 4–4), the main driving force behind plate tectonics. Convection is the motion within a fluid medium that results when temperatures differ at the bottom and the top. The core transfers heat to mantle rocks, whose increased buoyancy causes them to rise to the surface.

Convection currents and mantle plumes of hot rock transport molten magma to the underside of the lithosphere, which is responsible for most of the volcanic activity on the ocean floor and on the continents. Most mantle plumes originate from within the mantle, and some arise from the very bottom of the mantle, making the Earth's interior a huge bubbling pot stirred throughout its entire depth.

The formation of molten rock and the rise of magma to the surface results from an exchange of heat within the planet's interior. Fluid rocks in the mantle acquire heat from the core, ascend, dissipate heat to the lithosphere, cool, and descend back to the core where they are heated once again. The mantle currents travel very slowly, completing a single convection loop in several hundred million years.

The Earth is steadily losing heat from its interior to the surface through the lithosphere. About 70 percent of this heat loss results from seafloor spreading, while most of the rest is due to volcanism at subduction zones (Fig. 4–5).

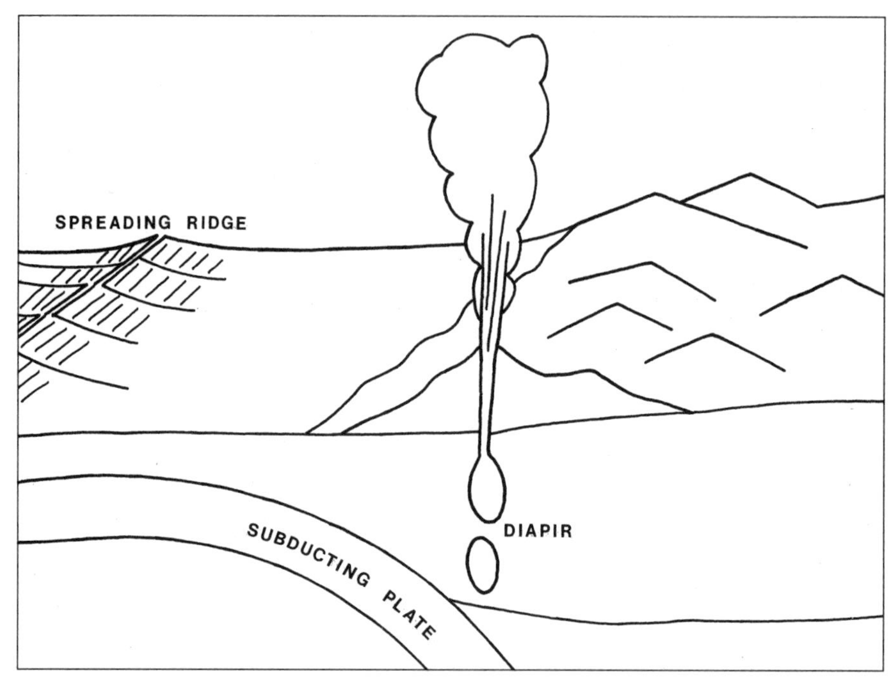

Figure 4–5 Volcanism at spreading ridges and subduction zones is responsible for most of the Earth's heat loss.

Lithospheric plates created at spreading ridges and destroyed at subduction zones are the final products of convection currents in the mantle.

Most of the mantle's heat originates from internal radiogenic sources. The rest comes from the core, which has retained much of its original heat since the early accretion of the Earth some 4.6 billion years ago. The temperature difference between the mantle and the core is nearly 1,000 degrees. Material from the mantle might be mixing with the fluid outer core to form a distinct layer on the surface that could block heat flowing from the core to the mantle and interfere with mantle convection.

The asthenosphere is the semi-molten region of the upper mantle upon which the rigid lithospheric plates ride. The asthenosphere is constantly losing material, which escapes from midocean ridges and adheres to the undersides of lithospheric plates. If the asthenosphere were not continuously fed new material from mantle plumes, the plates would grind to a complete halt, and the Earth would become, in all respects, a dead planet because all geologic activity would cease.

SEAFLOOR SPREADING

Seafloor spreading, which creates new lithosphere at spreading ridges on the ocean floor, begins with hot rocks rising from deeper portions of the mantle by convection currents. After reaching the underside of the lithosphere, the mantle rock spreads out laterally, dissipates heat near the surface, cools, and descends back into the deep interior of the Earth, where it receives more heat in a repeated cycle.

The constant pressure against the bottom of the lithosphere fractures the plate and weakens it. Convection currents flowing outward on either side of the fracture carry the separated parts of the lithosphere along with them, widening the gap in the plate. The rifting reduces the pressure in the underlying mantle, allowing mantle rocks to melt and rise through the fracture zone.

The molten rock passes through the lithosphere and forms magma chambers that supply molten rock for the generation of new lithosphere. Crustal material is sometimes introduced into the deep magma sources by subduction or off-scraping of a continental margin. The magma reservoirs resemble a mushroom up to 6 miles wide and 4 miles thick. The greater the supply of magma to the chambers, the higher the chambers elevate the overlying spreading ridge.

As magma flows outward from the trough between ridge crests, it adds new layers of basalt to both sides of the spreading ridge, creating new lithosphere. Some molten rock overflows onto the ocean floor in tremen-

Figure 4–6 Birth of a new Icelandic island, Surtsey, in November 1963, 7 miles south of Iceland. Courtesy of U.S. Navy

dous eruptions that generate additional oceanic crust. The continents ride passively on the lithospheric plates created at spreading ridges and destroyed at subduction zones. Therefore, the engine that drives the birth and evolution of rifts and, consequently, the breakup of continents and the formation of oceans, ultimately originates in the mantle.

The spreading ridges are the sites of frequent earthquakes and volcanic eruptions (Fig. 4–6). Over much of its length, the ridge system is carved down the middle by a sharp break or rift that is the center of an intense heat flow. Magma oozing out at spreading ridges erupts basaltic lava through long fissures in the trough between ridge crests and along lateral faults. The faults usually occur at the boundary between lithospheric plates, where the oceanic crust pulls apart by the plate separation. Magma welling up along the entire length of the fissure forms large lava pools that harden to seal the fracture.

The spreading ridge system is not a continuous mountain chain but broken into small, straight sections called spreading centers (Fig. 4–7). The movement of new lithosphere generated at the spreading centers produces a series of fracture zones, long, narrow linear regions up to 40 miles wide that consist of irregular ridges and valleys aligned in a stairstep shape. When lithospheric plates slide past each other as the seafloor spreads apart, they create transform faults ranging from a few miles to several hundred miles long. The transform faults transform from active faults between spreading ridge axes to inactive fracture zones past the ridge axes. The transform faults partition the midocean ridge system into independent segments, each with its own volcanic sources.

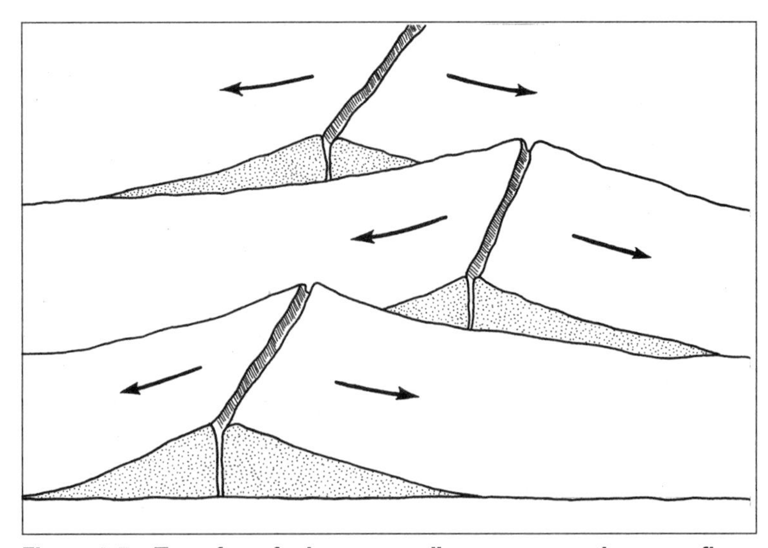

Figure 4–7 Transform faults at spreading centers on the ocean floor.

The transform faults of the Mid-Atlantic Ridge are offset laterally in a roughly east-west direction. The faults occur every 20 to 60 miles along the midocean ridge, where the longer offsets each consist of a deep trough joining the tips of two segments of the ridge. Friction between segments produces strong shearing forces, wrenching the ocean floor into steep canyons. Other types of offsets up to 15 miles wide separate several spreading centers, which are each 20 to 30 miles long. The end of one spreading center often runs past the end of another, and sometimes the tips of the segments bend toward each other.

Transform faults appear to result from lateral strain on the ocean floor, which is how rigid lithospheric plates are expected to react on the surface of a sphere. This activity is more intense in the Atlantic, where the spreading ridge system is steeper and more jagged than in the Pacific and Indian oceans. Transform faults dissecting the Mid-Atlantic Ridge generally are more rugged than those of the East Pacific Rise. Moreover, fewer widely spaced transform faults exist along the East Pacific Rise, where the rate of seafloor spreading is 5 to 10 times faster than at the Mid-Atlantic Ridge. Therefore, the crust affected by transform faults is younger, hotter, and less rigid in the Pacific than in the Atlantic, giving the Pacific undersea terrain much less relief.

BASALTIC MAGMA

The seafloor on the crest of the midocean ridge consists of hard volcanic rock. About 80 percent of all oceanic volcanism occurs along spreading ridges, where magma welling up from the mantle spews out onto the ocean floor. The spreading crustal plates grow by the steady accretion of solidifying magma. The molten magma beneath the spreading ridges consists mostly of peridotite, an iron-magnesium silicate. As the peridotite melts on its way through the lithosphere, a portion becomes highly fluid basalt. More than one square mile of new ocean crust, comprising about 5 cubic miles of basalt, forms throughout the world annually in this manner.

Magma rising toward the surface fills shallow reservoirs or feeder pipes that are the immediate source of volcanic activity. The magma chambers closest to the surface exist under spreading ridges, where the oceanic crust is only 6 miles or less thick. Large magma chambers lie under fast-spreading ridges where the lithosphere forms at a high rate, as in the Pacific. Narrow magma chambers lie under slow-spreading ridges such as those in the Atlantic.

As the magma chamber swells with molten rock and begins to expand, the crest of the spreading ridge bulges upward because of the buoyant forces generated by the magma. The greater the supply of molten magma, the higher it elevates the overlying ridge segment. The magma rises in narrow plumes that balloon out along the spreading ridge, upwelling as a passive

Figure 4–8 Pillow lava on Knight Island, Alaska. Photo by F. H. Moffit, courtesy of USGS

response to the release of pressure from plate divergence, somewhat like what happens when the lid is taken off a pressure cooker. Only the center of the plume is hot enough to rise all the way to the surface, however. If the entire plume erupted, it would build a massive volcano several miles high that would rival the tallest volcanoes found on other planets in the Solar System.

The main types of lava formations associated with midocean ridges are sheet flows and tube flows which form pillow lavas (Fig. 4–8). Sheet flows are more prevalent in the active volcanic zone of fast-spreading ridge segments like those of the East Pacific Rise, where in some places the plates separate at a rate of 5 or more inches per year. These flows consist of flat slabs of basalt usually less than a foot thick. The basalt that forms sheet flows is more fluid than that responsible for pillow structures. Pillow lavas often occur at slow-spreading ridges, such as the Mid-Atlantic Ridge, where plates separate at a rate of only about an inch per year and the lava is much more viscous. The manufacture of new oceanic crust in this manner explains why some of the most intriguing terrain features lie on the bottom of the ocean.

THE CIRCUM-PACIFIC BELT

Deep-sea trenches, where the ocean floor disappears into the Earth's interior, ring the Pacific Basin. Lithospheric plates descend sheetlike into

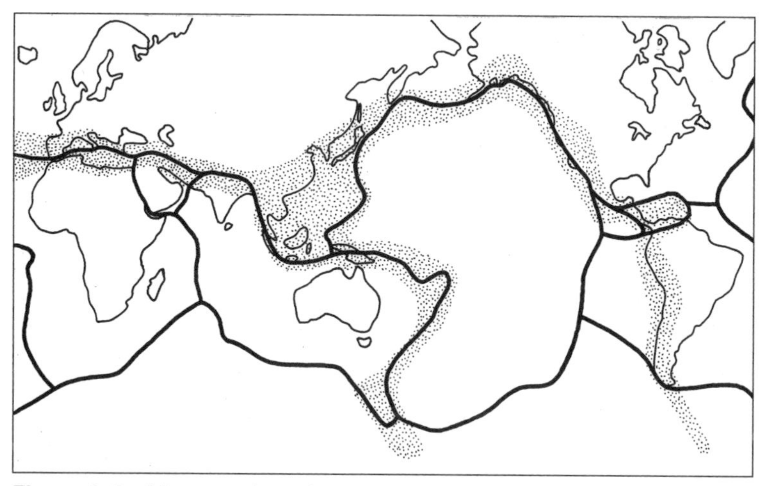

Figure 4–9 Most earthquakes occur in broad zones associated with plate boundaries.

the mantle at subduction zones, lying off continental margins and adjacent to island arcs. Plate subduction is responsible for the intense seismic activity that fringes the Pacific Ocean in a region known as the circum-Pacific belt, a chain of subduction zones flanking the Pacific Basin.

Most earthquakes originate at plate boundaries (Fig. 4–9). Wide bands of earthquakes mark continental plate margins, and narrow bands of earthquakes mark many major oceanic plate boundaries. The most powerful quakes are associated with plate subduction where one plate thrusts under another in deep subduction zones. The greatest amount of seismic energy occurs along the rim of the Pacific Ocean. In the western Pacific, the circum-Pacific belt encompasses volcanic island arcs that fringe the subduction zones, producing some of the largest earthquakes in the world.

The circum-Pacific belt is also known for extensive volcanic activity. Subduction zone volcanoes form island arcs, mostly in the Pacific, and most volcanic mountain ranges on the continents. The circum-Pacific belt coincides with the "ring of fire," which explains why the Pacific rim also contains the majority of the world's active volcanoes: the same tectonic forces that produce eartquakes are responsible for volcanic activity. The area of greatest seismicity is on the plate boundaries associated with deep trenches and volcanic island arcs, where an ocean plate dives under a continental plate.

Starting from New Zealand, which is divided by faults (Fig. 4–10), the circum-Pacific belt runs northward, encompassing the islands of Tonga, Samoa, Fiji, the Loyalty Islands, the New Hebrides, and the Solomons. The belt then runs westward to embrace New Britain, New Guinea, and the Moluccas Islands. One segment continues westward over Indonesia, while

Figure 4–10 Arrows indicate the Wellington Fault in New Zealand. Courtesy of USGS

the principal arm travels northward to encompass the Philippines, where a large fault zone runs from one end of the islands to the other. The seismic belt continues on to Taiwan and the Japanese archipelago, which has been hard hit by major earthquakes.

An inner belt runs parallel to the main belt and takes in the Marianas, a string of volcanic islands characterized by a massive trench system in places over 30,000 feet deep. The belt continues northward and follows the seismic arc across the top of the Pacific, comprising the Kuril Islands (devastated by an 8.2 magnitude earthquake on October 4, 1994), the Kamchatka Peninsula, and the Aleutian Islands, which constantly rock and roll. The Aleutian Trench, the largest on Earth, is responsible for the many great earthquakes that strike Alaska. A 200-mile-long stretch called the Shumagin gap, which is accumulating huge stresses in the descending Pacific plate, is poised for a massive earthquake.

Crossing over to the eastern side of Pacific Basin, the seismic belt continues along the Cascadia subduction zone, extending along the coast from southern British Columbia to northern California. This belt has severely shaken the Pacific Northwest in the geologic past and is responsible for numerous powerful volcanoes. The tectonic activity is generated by the Juan de Fuca and Gorda plates slipping under the North American plate. The San Andreas Fault (Fig. 4–11), which marks the boundary between the North American and Pacific plates, rattles much of California.

The Andes Mountain regions of Central and South America, especially in Chile and Peru, have been lashed by some of the world's strongest and most destructive earthquakes. The 1960 Chilian earthquake of 9.5 magnitude, the largest in modern history, elevated a California-size chunk of crust some 30 feet. In this century alone, some two dozen earthquakes of 7.5 magnitude or greater have devastated the region.

An immense subduction zone lying just off the coast influences the whole western seaboard of South America. The lithospheric plate on which the South American continent rides forces the Nazca plate to buckle under, causing great tensions to build deep within the crust. While some rocks shove downward, others thrust to the surface to raise the Andes Mountains, the fastest growing mountain range on Earth. The resulting forces build great stresses into the entire region. As the strain builds and the crust cracks, great earthquakes roll across the countryside.

Figure 4–11 View south along the San Andreas Fault in the Carrizo Plains, California. Photo by R. E. Wallace, courtesy of USGS

TABLE 4–1 THE WORLD'S OCEAN TRENCHES

Trench	Depth (miles)	Width (miles)	Length (miles)
Peru-Chile	5.0	62	3700
Java	4.7	50	2800
Aleutian	4.8	31	2300
Middle America	4.2	25	1700
Mariana	6.8	43	1600
Kuril-Kamchatka	6.5	74	1400
Puerto Rico	5.2	74	960
South Sandwich	5.2	56	900
Philippine	6.5	37	870
Tonga	6.7	34	870
Japan	5.2	62	500

THE DEEP-SEA TRENCHES

The creation of new lithosphere at midocean ridges is matched by the destruction of old lithosphere at subduction zones (Fig. 4–12). Deep trenches lying at the edges of continents or along volcanic island arcs mark the seaward boundaries of the subduction zones. As a lithospheric plate sinks into the mantle, the line of subduction creates a deep-sea trench. While the Pacific plate drifts toward the northwest, its leading edge dives into the mantle, forming the deepest trenches in the world. The Mariana Trench in the western Pacific is the lowest point on Earth. It extends northward from the Island of Guam in the Mariana Islands and reaches a depth of nearly 7 miles below sea level.

Subduction zones, where cool, dense lithospheric plates dive into the mantle, are regions of low heat flow and high gravity (an area where the gravitational pull is strong relative to the average force of gravity on the earth's surface). Conversely, because of their extensive volcanic activity, the associated island arcs are regions of high heat flow and low gravity. The deep-sea trenches are regions of intense volcanism, producing the most explosive volcanoes on Earth. Volcanic island arcs, which typically share similar curved shapes and similar volcanic origins, fringe the trenches. These island chains, for example the Aleutian Islands and the islands of Japan, are generally arc-shaped because of the geometry of the ocean floor. Any time a plane (in this case a rigid lithospheric plate) cuts into a sphere (here the mantle-encrusted Earth), the point of intersection forms an arc.

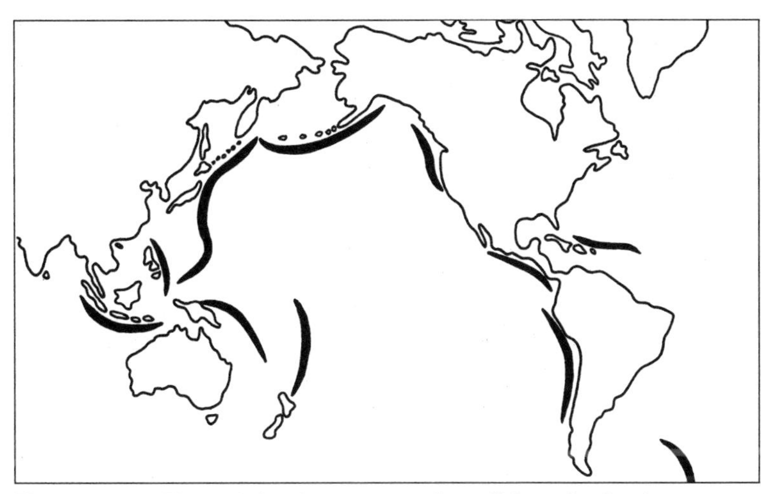

Figure 4–12 The subduction zones where lithospheric plates enter the mantle are marked by the deepest trenches in the world.

Think of a knife (plane) slicing into a cantaloupe (sphere) and the arced smile it produces.

The trenches are also sites of almost continuous earthquake activity deep in the bowels of the Earth, about 2 miles down. Plate subduction causes stresses to build into the descending lithosphere, producing deep-seated earthquakes that outline the boundaries of the plate. A band of recent shallow earthquakes clustered in a line running through Microneasia might

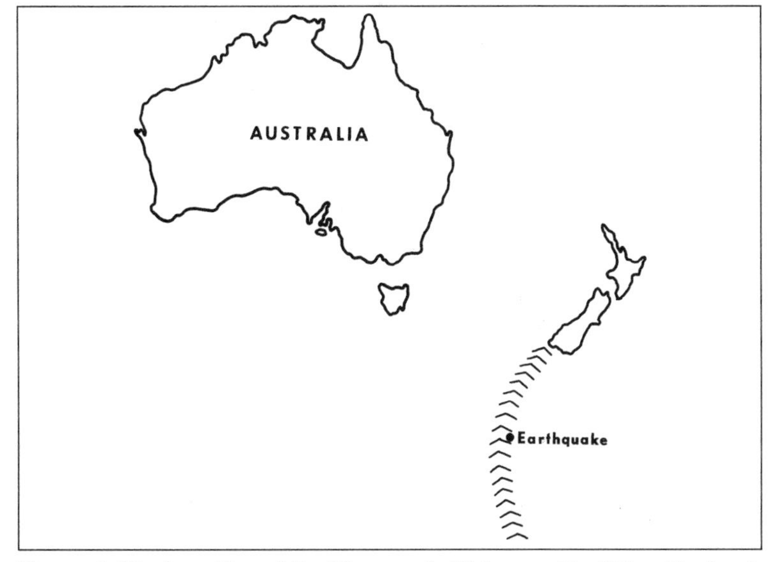

Figure 4–13 Location of the Macquarie Ridge south of New Zealand.

mark the earliest stages in the birth of a subduction zone and indicate the formation of a trench to the north and west of New Guinea. Gravity in the area is lower than normal—a sign of a trench caused by the sagging of the ocean floor. In addition, a bulge in the crust to the south of this area suggests that the edge of a slab of crust is beginning to dive into the Earth. The subduction process might not be operating fully for another 5 or 10 million years as the deep-sea trench nibbles away at the Pacific plate.

The seafloor south of New Zealand could also be experiencing the early stages of subduction in the process of creating a deep-sea trench. A geologic scar on the floor of the Pacific known as the Macquarie Ridge is still evolving as part of this process. The ridge is an undersea chain of mountains and troughs that runs south from New Zealand (Fig. 4–13) and forms the boundary between the Australian and Pacific plates, which are moving past each other in opposite directions. In 1989, a massive earthquake of 8.2 magnitude struck the ridge.

As the Australian plate slides northwest past the Pacific plate, ruptures occur along vertical faults between the plates, creating large strike-slip earthquakes. As they pass one another, the plates are also pressing together along dipping fault planes, creating smaller compressional earthquakes. This suggests that subduction is just beginning along the Macquarie Ridge. However, the separate dipping faults that flank the area have not yet connected to form a single large fault plane, a necessary first step before full-fledged subduction commences.

A plate extending away from its place of origin at a midocean spreading ridge thickens and becomes denser as additional material from the asthenosphere adheres to its underside in a process called underplating. The depth at which a lithospheric plate sinks as it moves away from a spreading ridge increases with age. Thus, the older the lithosphere, the more basalt that underplates it, making the plate thicker, denser, and deeper.

Eventually, the plate becomes so dense that it loses buoyancy and sinks into the mantle, and the subduction creates a deep-sea trench at clearly defined subduction zones. As the subducted portion of the plate dives into the Earth's interior, the rest of the plate, which might carry a continent on its back, is pulled along with it like a freight train hauled by a locomotive. The force of the pull on the sinking slab depends on the length of the subduction zone, the rate of subduction, and the amount of trench suction produced by mantle convection. Plate subduction is the main driving force behind plate tectonics, and pull at subduction zones is more active than push at spreading ridges to move the continents around the surface of the globe.

New oceanic crust generated by seafloor spreading in the Atlantic and the eastern Pacific is offset by the subduction of old oceanic crust along the rim of the Pacific to make more room. Because the seafloor spreading rate is not always the same as the rate of subduction, associated midocean ridges often move laterally. Most of the subduction zones are in the western

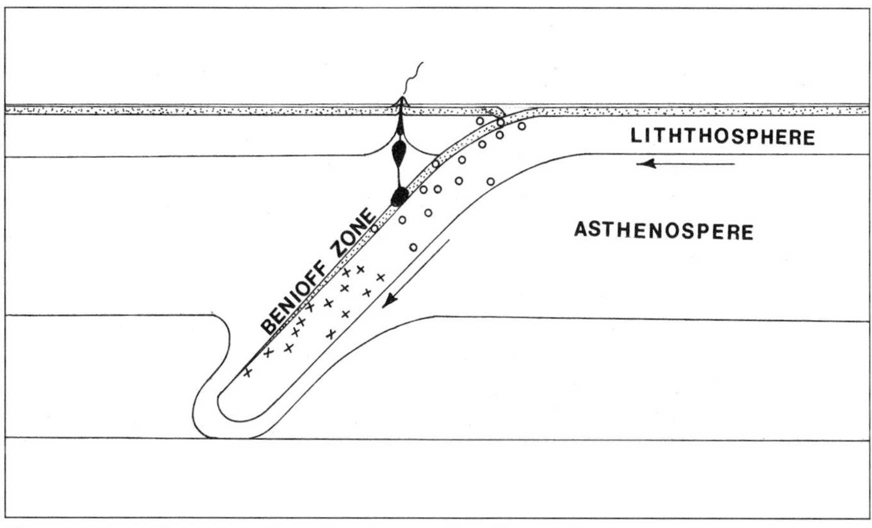

Figure 4–14 Cross-section of a descending lithospheric plate. Each *O* denotes a shallow earthquake. Each *X* denotes a deep-seated earthquake. The Benioff zone is a plane of seismic activity, marking the outline of a descending plate.

Pacific, which accounts for the fact that no oceanic crust is older than 170 million years.

Subduction zones are the sites of almost continuous seismic activity, with a band of earthquakes marking the boundaries of a sinking lithospheric plate (Fig. 4–14). As plates slide past each other along subduction zones, they create highly destructive earthquakes, such as those that have always plagued Japan, the Philippines, and other islands connected with subduction zones.

The subduction zones are also regions of intense volcanic activity, producing the most explosive volcanoes on the planet. Magma reaching the surface of the oceanic crust erupts on the ocean floor, creating new volcanic islands. Most volcanoes do not rise above sea level; rather, they become isolated undersea volcanic structures called seamounts. The Pacific Basin is more volcanically active and has a higher density of seamounts than the Atlantic or Indian basins. Subduction zone volcanoes are so explosive because their magmas contain large quantities of volatiles and gases that escape violently when reaching the surface. The type of volcanic rock erupted in this manner is andesite, named for the Andes Mountains that form the spine of South America and that are well known for their violent eruptions.

PLATE SUBDUCTION

As the rigid lithospheric plate carrying the oceanic crust descends into the Earth's interior, it slowly breaks up and melts. Over a period of millions of years, it is absorbed into the general circulation of the mantle. When the plate dives into the interior, most of its trapped water goes down with it becoming an important volatile in magma. The subducted plate also supplies molten magma for volcanoes, most of which ring the Pacific Ocean and recycle chemical elements to the Earth.

The amount of subducted plate material is vast. When the Atlantic and Indian oceans opened up and began forming new oceanic crust some 125 million years ago, an equal area of oceanic crust disappeared into the mantle. This meant that 5 billion cubic miles of crustal and lithospheric material was destroyed. At the present rate of subduction, the mantle will consume an area equal to the entire surface of the planet in 160 million years.

The convergence of lithospheric plates forces the thinner, more dense oceanic plate under the thicker, more buoyant continental plate. When oceanic plates collide, the older and denser plate dives under the younger plate (Fig. 4–15). A deep-ocean trench marks the line of initial subduction. At first the plate's angle of descent is low, but it gradually steepens to about 45 degrees, with the rate of vertical descent (typically 2 to 3 inches per year) less than the rate of horizontal motion of the plate.

If continental crust moves into a subduction zone, its greater buoyancy prevents it from being dragged down into the trench. When two continental plates converge, the crust is scraped off the subducting plate and fastens onto the overriding plate, welding the two pieces of continental crust together. Meanwhile, the subducted lithospheric plate, now without its

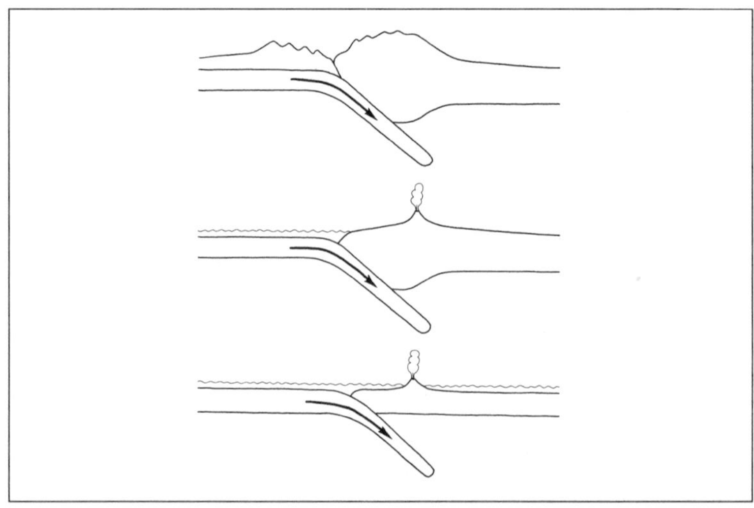

Figure 4–15 Collision between two continental plates (top), a continental plate and an oceanic plate (middle), and two oceanic plates (bottom).

overlying crust, continues to dive into the mantle, squeezing the continental crusts together and forcing up mountain ranges.

In many subduction zones, such as the Lesser Antilles, sediments and their contained fluids are removed by offscraping and underplating in accretionary prisms, wedges of sediment that form on the overriding plate adjacent to the trench. In other subduction zones, such as the Mariana and Japan trenches, little or no sediment accretion occurs. Thus, subduction zones differ markedly from one another in the amount of sedimentary material removed at the accretionary prism. In most cases at least some sediment and bound fluids appear to be subducted to deeper levels.

The underthrusting of continental crust by additional crustal material increases continental buoyancy and pushes up mountain ranges. A similar process occurred when India collided with Asia about 45 million years ago, pushing up the Himalayas. A strange series of east-west wrinkles in the ocean crust just south of India verifies that the Indian plate is still pushing northward, shrinking the Asian continent as much as 3 inches a year. Further compression and deformation might eventually take place beyond the line of collision, producing a high plateau with surface volcanoes, similar to the Tibetan Plateau, the largest in the world.

When continental and oceanic plates converage, the denser oceanic plate dives beneath the lighter continental plate and is forced farther downward. The sedimentary layers of both plates are squeezed like an accordion, swelling the leading edge of the continental crust to create folded mountain belts such as the Appalachians. As the descending plate dives farther under the continent, it reaches depths where the temperatures are extremely high. The upper part of the plate melts to form magma that rises toward the surface to provide volcanoes with a new supply of molten rock.

5

SUBMARINE VOLCANOES

An extraordinary number of volcanoes are hidden under the waves. More than 80 percent of the Earth's surface above and below the sea is of volcanic origin. The vast majority of the volcanic activity that continually remakes the surface of the Earth takes place at the bottom of the ocean, where the most of the world's volcanoes are located. Oceanic volcanoes also happen to be among the most explosive in the world, and whole islands have been known to disappear, with new ones popping up to take their places.

Nearly all the world's islands started out as undersea volcanoes. In volcanic island-building, successive eruptions pile up volcanic rocks until the volcano's peak finally breaks the surface of the sea. Active undersea volcanoes rising tens of thousands of feet off the ocean floor become volcanic islands. Measured from the seabed, some of them are the world's tallest mountains.

THE RING OF FIRE

Most of the world's volcanoes are associated with crustal movements at the margins of lithospheric plates. An almost continuous "ring of fire" runs

TABLE 5–1 COMPARISON BETWEEN TYPES OF VOLCANISM

Characteristic	Subduction	Rift Zone	Hot Spot
Location	Deep ocean trenches	Midocean ridges	Interior of plates
Percent active volcanoes	80%	15%	5%
Topography	Mountains, island arcs	Submarine ridges	Mountains, geysers
Examples	Andes Mts. Japan Is.	Azores Is. Iceland	Hawaiian Is. Yellowstone
Heat source	Plate friction	Convection currents	Upwelling from core
Magma temperature	Low	High	Low
Magma viscosity	High	Low	Low
Volitile content	High	Low	Low
Silica content	High	Low	Low
Type of eruption	Explosive	Effusive	Both
Volcanic products	Pyroclasts	Lava	Both
Rock type	Rhyolite Andesite	Basalt	Basalt
Cone type	Composite	Cinder fissure	Cinder shield

along the rim of the Pacific, which coincides with the circum-Pacific belt because the same tectonic processes that generate earthquakes also produce volcanoes. The greatest activity occurs on plate boundaries associated with deep trenches along volcanic island arcs and the margins of continents.

The Ring of Fire corresponds to a band of subduction zones surrounding the Pacific Basin (Fig. 5–1), which have devoured almost all the seafloor since the breakup of Pangaea. The oldest oceanic crust lies in a small patch off southeast Japan and is only about 170 million years old, compared with the ocean floor, which is on average about 100 million years. While subducting into the mantle, the oceanic crust melts to provide molten magma for volcanoes that fringe the deep-sea trenches. This is why most of the 600 active volcanoes in the world lie in the Pacific Ocean—nearly half of them in the western Pacific region alone.

Subduction zone volcanism builds volcanic chains on the continents and island arcs in the ocean. At convergent plate boundaries, where one plate subducts under another, magma forms when the lighter constituents of the subducted oceanic crust melt. The upwelling magma creates island arcs,

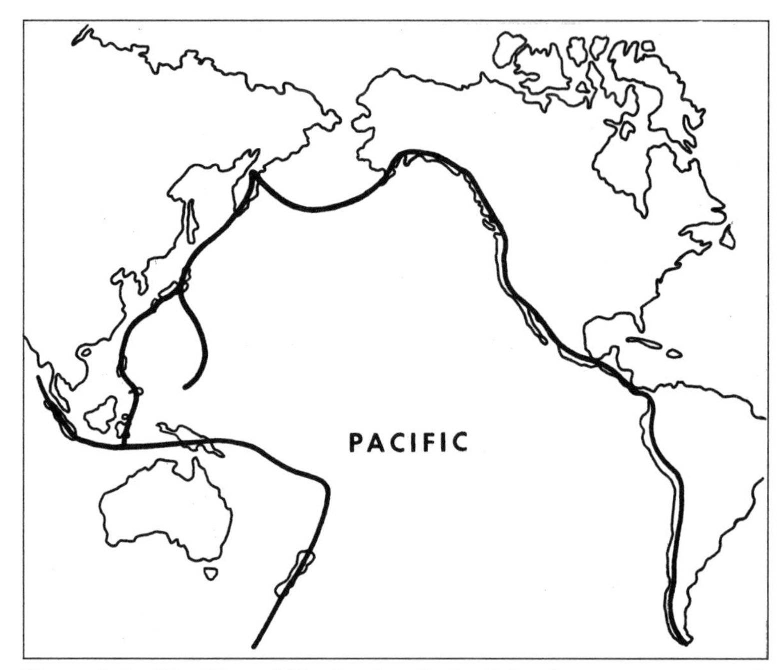

Figure 5–1 The Ring of Fire is a band of subduction zones surrounding the Pacific Ocean.

including Indonesia, the Philippines, Japan, the Kuril Islands, and the Aleutians, the longest, extending more than 3,000 miles from Alaska to Asia.

Beginning at the western tip of the Aleutian Islands off Alaska, the Ring of Fire runs along the Aleutian archipelago, a string of volcanic islands (Fig. 5–2) created by the subduction of the Pacific plate down the Aleutian

Figure 5–2 A crater and dome of Great Sitkin Volcano, Great Sitkin Island, Aleutian Islands, Alaska. Photo by F. S. Simons, courtesy of USGS

Trench. The band of volcanoes turns south across the Cascade Range of British Columbia, Washington, Oregon, and northern California, associated with the subduction of the Juan de Fuca plate down the Cascadia subduction zone. The ring then runs across Baja California and southwest Mexico, where lie the volcanoes Parícutin and El Chichon. (The latter is perhaps the dirtiest volcano of this century in terms of ash cast up into the atmosphere.)

The volcano belt continues through western Central America, which has numerous active cones, including Nevado del Ruiz of Columbia, whose destructive mudflows killed 25,000 people in November 1985. It is among some 20 other volcanoes that have each killed more than 1,000 people since 1700 (Fig. 5–3). The Ring of Fire journeys along the course of the Andes Mountains on the western edge of South America, whose highly explosive nature results from the subduction of the Nazca plate down the Chilean Trench.

The volcanic band then turns toward Antarctica and the islands of New Zealand, New Guinea, and Indonesia, where the volcanoes Tambora and Krakatoa produced the greatest eruptions in modern history. These eruptions were instigated by the subduction of the Australian plate down the Java Trench. The band continues across the Philippines, where Mount Pinatubo spewed a massive eruption cloud in June 1991 that caused dramatic changes in climate; this eruption resulted from the subduction of the Pacific plate down the Philippine Trench. The Ring of Fire runs across Japan (where the Fuji volcano reigns majestically), finally ending on the Kamchatka Peninsula in northeast Asia.

Subduction zone volcanoes such as those in the western Pacific and Indonesia (Figs. 5–4a&b) are among the most explosive in the world,

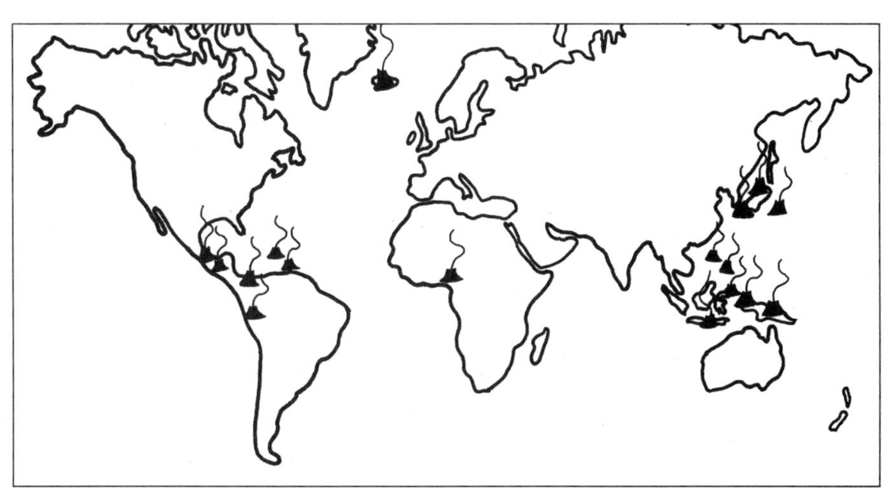

Figure 5–3 Locations of killer volcanoes responsible for the deaths of over 1,000 people each since 1700.

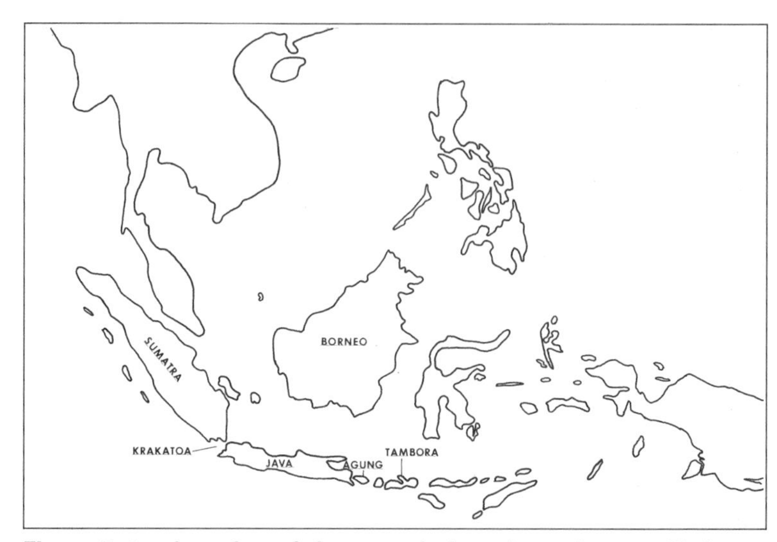

Figure 5–4a Location of the great Indonesian volcanoes Krakatoa and Tambora.

Figure 5–4b An active volcano on Andonara Island, Indonesia (center), leaves a 30-mile-long train of ash. Courtesy of NASA

destroying entire islands when they erupt. One classic example is the near-total destruction of the Indoneasian island of Krakatoa in 1883. The explosive nature of such volcanoes is due to abundant silica and volatiles in the magma, which consists of water and gases derived from sediments on the ocean floor that were subducted into the mantle and melted. When the magma depressurizes as it reaches the surface, the volatiles explode and fracture the molten rock, destroying much of the volcano in the process.

On the continents, plate subduction creates long chains of powerful volcanoes. The Cascade Range in the Pacific Northwest is a belt of volcanoes associated with a subduction zone under the North American continent. The Andes Mountains of South America are a chain of volcanoes associated with a subduction zone under the South American continent. As the lithosphere plunges into the mantle, tremendous heat melts the descending plate and the adjacent lithospheric plate, and magma rising to the surface feeds rows of active volcanoes.

THE RISING MAGMA

Subduction zones created by descending plates accumulate large quantities of sediment from the adjacent continents and volcanic island arcs. When the sediments and seawater become trapped between a subducting oceanic

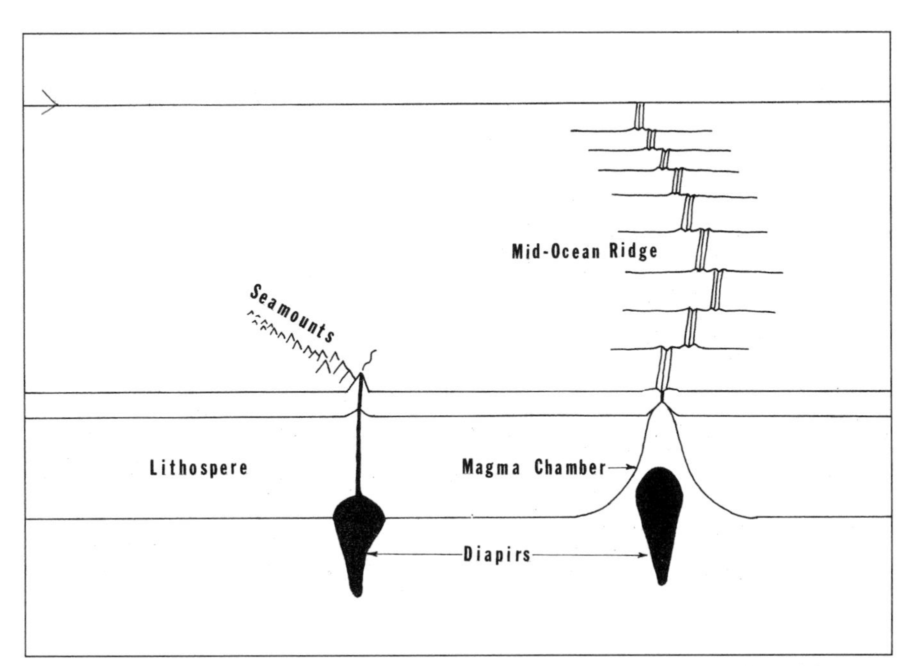

Figure 5–5 Diapirs supply the magma for volcanoes and spreading ridges on the ocean floor.

plate and an overriding continental plate, they undergo strong deformation, shearing, heating, and metamorphism (recrystallization without melting). The sediments are carried deep into the mantle, where they melt to become the source of new magma for volcanoes fringing the subduction zones.

Some magma originates from the partial melting of subducted oceanic crust, with heat supplied by the shearing action at the top of the descending plate. Convective motions in the wedge of asthenosphere caught between the descending oceanic plate and the continental plate forces material upward, where it melts under reduced pressures.

The magma rises to the surface in giant plumes called diapirs. Upon reaching the underside of the lithosphere, the diapirs burn holes through the crust as the molten rock melts its way upward. As the diapirs rise toward the surface, they form magma bodies, which become the immediate source for new igneous activity (Fig. 5–5). After reaching the ocean floor, the magma erupts to create new volcanic islands.

The rock type associated with subduction zone volcanoes is fine-grained gray andesite, which contains abundant silica from deep-seated sources, possibly 70 miles below the surface. The rock derives its named from the Andes Mountains, whose volcanoes are highly explosive because of large

TABLE 5–2 CLASSIFICATION OF VOLCANIC ROCKS

Property	Basalt	Andesite	Rhyolite
Silica content	Lowest about 50%, basic rock	Intermediate about 60%	Highest more than 65%, acid rock
Dark mineral content	Highest	Intermediate	Lowest
Typical minerals	Feldspar Pyroxene Olivine Oxides	Feldspar Amphibole Pyroxine Mica	Feldspar Quartz Mica Amphibole
Density	Highest	Intermediate	Lowest
Melting point	Highest	Intermediate	Lowest
Molten rock viscosity at the surface	Lowest	Intermediate	Highest
Tendency to form lavas	Highest	Intermediate	Lowest
Tendency to form pyroclastics	Lowest	Intermediate	Highest

Figure 5–6 A submarine eruption of Myojin-sho Volcano in the Izu Islands, Japan, on September 23, 1952. Courtesy of USGS

amounts of volatiles in the magma. As the magma rises toward the surface, the pressure drops and volatiles escape with great force, shooting out of the volcano like pellets fired from a gigantic canon.

The mantle material that slowly extrudes onto the surface is black basalt, the most common volcanic rock. The ocean floor is paved with abundant basalt, and most volcanoes are entirely or predominately basaltic. The magma that forms basalt originates in a zone of partial melting in the upper mantle more than 60 miles below the surface. The semimolten rock at this depth is less dense and more buoyant than the surrounding mantle material and rises slowly toward the surface.

As the magma ascends, the pressure decreases, allowing more mantle material to melt. Volatiles such as dissolved water and gases make the magma flow easily. The mantle material below spreading ridges that create new oceanic crust consists mostly of peridotite, which is rich in silicates of iron and magnesium. As the peridotite melts on its journey to the surface, a portion becomes highly fluid basalt.

The magma's composition indicates both its source materials and the depth within the mantle from which they originated. The degree of partial melting of mantle rocks, partial crystallization that enriches the melt with silica, and the assimilation of a variety of crustal rocks in the mantle all influence the composition of the magma. When the erupting magma rises toward the surface, it incorporates a variety of rock types along the way, which also changes its composition. The magma's composition determines its viscosity and the type of eruption that will occur. If the magma is highly fluid and contains little dissolved gas upon reaching the surface, it produces basaltic lava, and the eruption is usually quite mild, as with the Hawaiian volcanoes. If, however, the magma rising toward the surface contains a large quantity of dissolved gases, the eruption can be highly explosive and very destructive. Water is possibly the single most important volatile in magma and affects the explosive nature of some volcanic eruptions by causing a rapid expansion of steam as the magma reaches the surface (Fig. 5–6).

ISLAND ARCS

Almost all volcanic activity is confined to the margins of lithospheric plates. As previously noted, deep trenches at the edges of continents or along volcanic island arcs mark the seaward boundaries of subduction zones. At convergent plate boundaries, where one plate subducts under another, new magma forms when the lighter constituent of the subducted plate melts and rises to the surface. When the upwelling magma erupts on the ocean floor, it creates island arcs. These occur mostly in the Pacific.

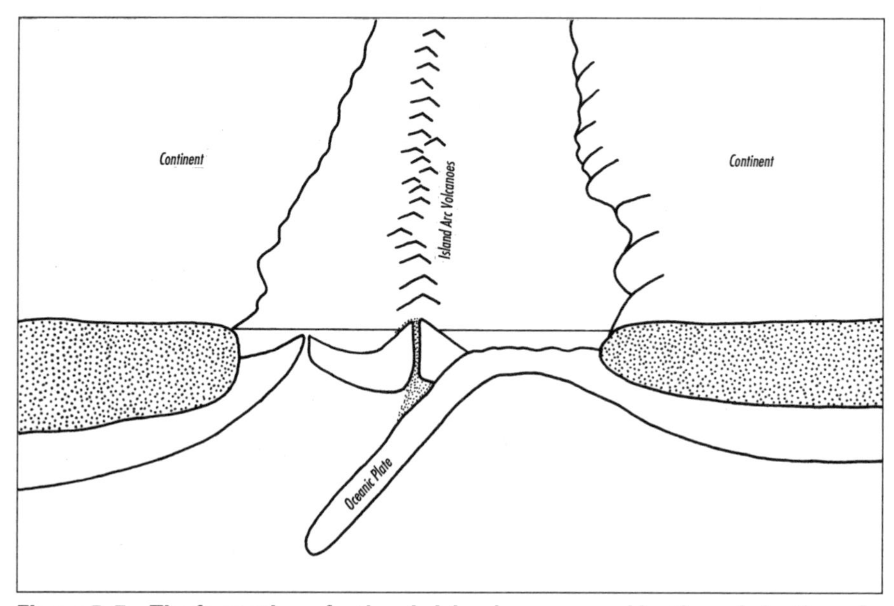

Figure 5–7 The formation of volcanic island arcs caused by the subduction of a lithospheric plate.

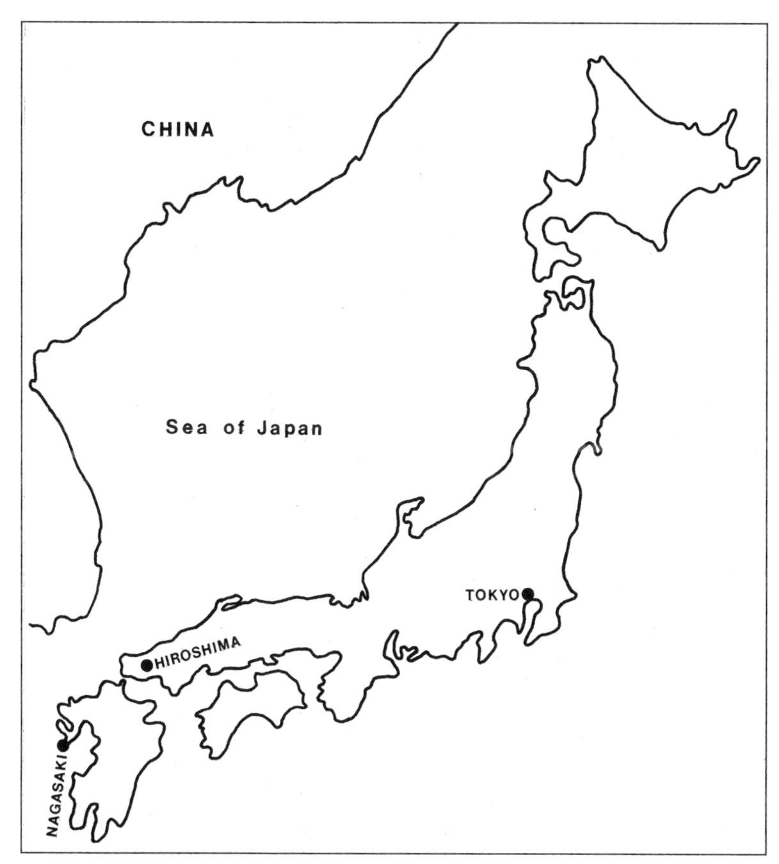

Figure 5–8 Location of the Sea of Japan.

The longest island arc is the Aleutian Islands, extending more than 3,000 miles from Alaska to Asia, where the Pacific plate subducts beneath the overriding North American plate. The Kurile Islands to the south form another long arc. The islands of Japan, Philippines, Indonesia, New Hebrides, Tonga, and those from Timor to Sumatra also form island arcs. These island arcs are all similarly curved, have similar geological compositions, and are associated with subduction zones. The curvature of the island arcs results from the curvature of the Earth. Just as an arc forms when a plane cuts a sphere, so does an arc-shaped feature result when a rigid lithospheric plate subducts into the Earth's spherical mantle.

At deep-sea trenches, created during the subduction process, magma forms when oceanic crust thrust deep into the mantle melts. As the lithospheric plate carrying the oceanic crust descends farther into the Earth's interior, it slowly breaks up and also melts. Over a period of millions of years, it assimilates into the general circulation of the mantle, possibly descending as deep as the top of the Earth's core. Eventually, the magma rises to the surface in giant plumes, completing the loop in the convection of the mantle.

The subducted plate is also the immediate supply of molten magma for volcanic island arcs (Fig. 5–7). Behind each island arc is a marginal or a back-arc basin, a depression in the ocean crust caused by plate subduction. Steep subduction zones like the Mariana Trench in the western Pacific form back-arc basins, whereas shallow ones like the Chilean Trench off the west coast of South America do not. A classic back-arc basin is the Sea of Japan between China and the Japanese archipelago (Fig. 5–8); the archipelago is comprised of ruptured continental fragments. Gradually, the sea will close off entirely as the Japanese islands slam into Asia.

Back-arc basins are regions of high heat flow because they overlie relatively hot material brought up by convection currents behind the island arcs or by upwelling from deeper regions in the mantle. The trenches are regions of low heat flow because of the subduction of cool dense lithosphere, while the adjacent island arcs generally are regions of high heat flow because of their high degree of volcanism.

GUYOTS AND SEAMOUNTS

Marine volcanoes associated with midocean ridges that rise above the sea become volcanic islands. Most of the world's islands began as undersea volcanoes. Successive volcanic eruptions pile up layers of basalt until the peak finally breaks through the ocean surface. The volcanic ash makes a rich soil; as the island cools, seeds carried by wind, sea, and animals rapidly

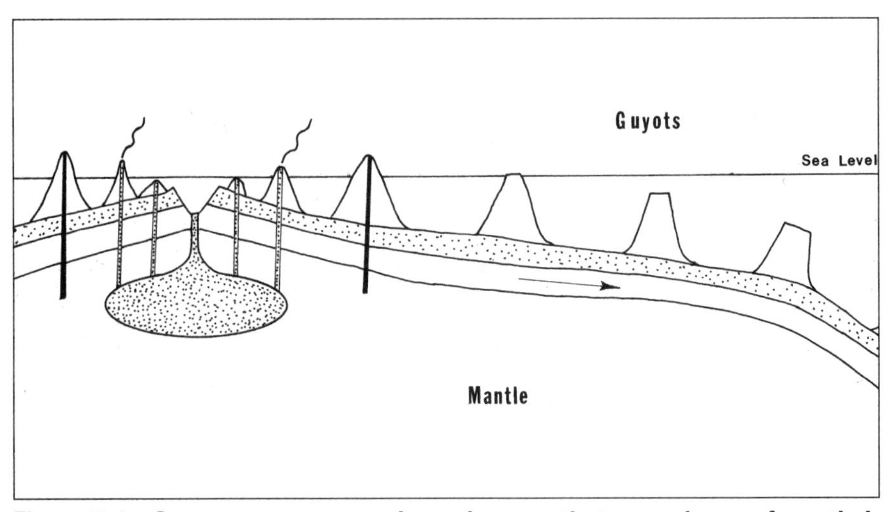

Figure 5–9 Guyots were once active volcanoes that moved away from their magma source and have since disappeared beneath the sea.

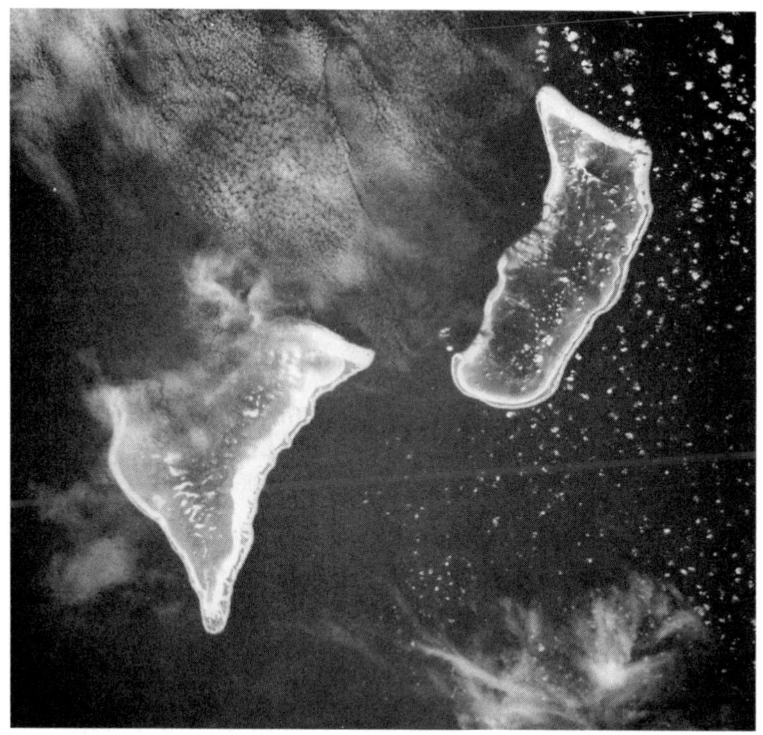

Figure 5–10 Tarawa (left) and Abaiang Atolls, Gilbert Islands. Courtesy of NASA

turn the newly formed land into a lush tropical paradise. Life must still cope with the rumblings deep within the Earth, however, and the island eventually might be destroyed in a single huge convulsion.

Most volcanic islands, however, end their lives quietly, eroded by the incessant pounding of the sea. Submarine volcanoes called guyots (pronounced "ghee-ohs") located in the Pacific once towered above the ocean. But the constant wave action eroded them below the sea surface, leaving them as though the tops of the cones had been sawed off. The farther these volcanoes were conveyed from volcanically active regions, the older and flatter they became (Fig. 5–9), suggesting that the guyots and the plates they rode on drifted across the ocean floor far from their places of origin. The islands appeared to have formed in assembly-line fashion, each moving in succession away from a magma chamber lying beneath the ocean floor.

Beyond the oldest Hawaiian island, Kauai, the persistent pounding of the waves has eroded the ancient volcanoes so that they now lie below sea level. Coral atolls, like Midway Island, and shallow shoals were formed by coral living on the flattened tops of eroded volcanoes. Atolls are rings of coral islands, enclosing a central lagoon (Fig. 5–10). They consist of coral reefs up to several miles across, and many atolls formed on ancient volcanic cones that have subsided beneath the sea, with the rate of coral growth matching the rate of subsidence. Continuing in a northwestward direction is an associated chain of undersea volcanoes, called the Emperor Seamounts (Fig. 5–11); these were presumably built by a single hot spot, although how such a plume could persist for over 70 million years remains a mystery.

Most marine volcanoes never rise above the sea to become islands, but instead become isolated undersea volcanoes called seamounts. Magma upwelling from the upper mantle at depths of more than 60 miles below the surface concentrates in narrow conduits that lead to a main feeder column. The magma erupts on the ocean floor, building seamounts, which are generally isolated and strung out in chains across the interior of a plate. Some seamounts are associated with extended fissures, along which magma wells up through a main conduit, piling successive lava flows on one another.

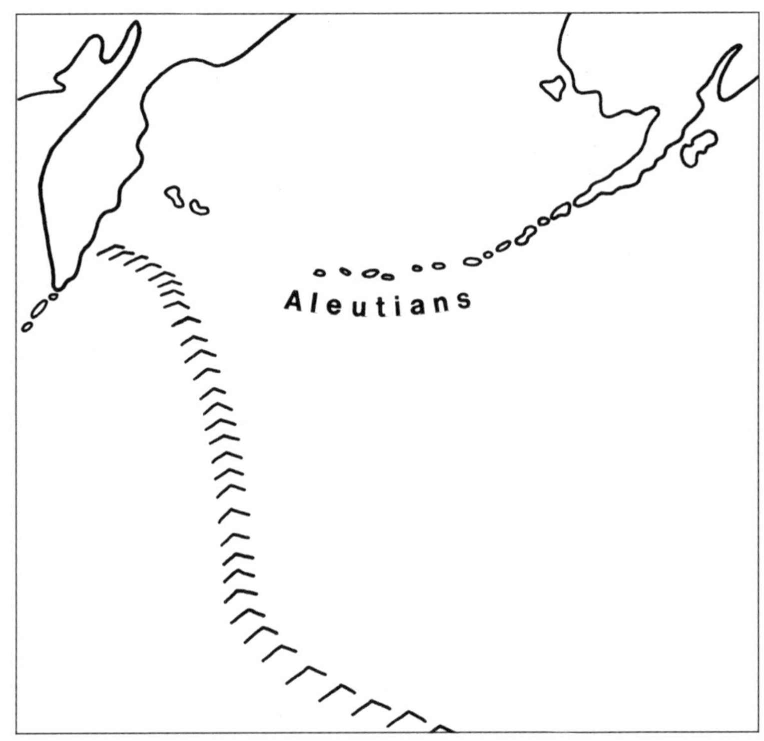

Figure 5–11 The Emperor Seamounts and Hawaiian Islands in the North Pacific represent motions of the Pacific plate over a volcanic hot spot. Note the sharp northward bend in the seamounts, caused by shifting of the Pacific plate.

Figure 5–12 A broad fountain pit in the cinder cone and large lava rivers draining from it, Halemaumau Volcano, Hawaiian Islands. Photo by G. A. MacDonald, courtesy of USGS

The crust under the Pacific Ocean is more volcanically active than the Atlantic or Indian oceans, providing a higher density of seamounts. The tallest seamounts rise over 2.5 miles above the seafloor in the western Pacific near the Philippine Trench. The number of undersea volcanoes increases with advanced crustal age and increased thickness. Deep-sea ridges called abyssal hills, which were developed by eruptions along midocean ridges, cover 60 to 70 percent of the Earth's surface. The average density of Pacific seamounts is 5 to 10 volcanoes per 5,000 square miles of ocean floor, by far outnumbering volcanoes on the continents.

Sometimes the summit of a seamount contains a crater, within which lava extrudes. If the crater exceeds a mile in diameter it is called a caldera, whose depth below the crater rim can be as much as 1,000 feet. Calderas form when the magma reservoir empties, creating a hollow chamber. Without support, the top of the volcanic cone collapses, forming a wide depression similar to calderas of Hawaiian volcanoes (Fig. 5–12). Feeder vents along the periphery of the caldera supply fresh lava that fills the caldera, giving the volcano a flattop appearance. Other undersea volcanoes do not have a collapsed caldera; instead, the summit contains several isolated volcanic peaks rising upward of 1,000 feet high.

RIFT VOLCANOES

More than three-quarters of all oceanic volcanism occurs at midocean ridges, where basaltic magma wells up from the mantle and spews out onto the ocean floor in response to seafloor spreading. Lithospheric plates subduct into the mantle like great sheets and arise again in giant cylindrical plumes of hot rock at midocean ridges. A series of plumes miles apart feed separate segments of the spreading ridge.

At the crest of a midocean ridge, the ocean floor consists almost entirely of hard volcanic rock. Along much of its length, the ridge system is divided down the middle by a sharp break or rift that is the center of volcanic activity. The spreading ridges are the sites of frequent earthquakes and volcanic eruptions, as though the entire system were a series of giant cracks in the crust, from which molten magma oozes out onto the ocean floor.

The volcanic activity associated with midocean rift systems is fissure eruption, the most common type. Such eruptions build typical conical volcanic structures. Volcanoes formed on or near midocean ridges often develop into isolated peaks when they move away from the ridge axis as the seafloor spreads apart. During fissure eruptions, lava bleeds through fissures in the trough between ridge crests and along lateral faults. The faults usually occur at the boundary between lithospheric plates, where the oceanic crust splits apart by the separating plates. Magma welling up along the entire length of the fissure forms large lava pools, similar to those of broad shield volcanoes such as the Hawaiian volcano Mauna Loa, the largest of its kind in the world.

Seamounts associated with midocean ridges that grow tall enough to break through the surface of the ocean become volcanic islands. The Galapagos Islands west of Ecuador are volcanic islands associated with the East Pacific Rise. Volcanic islands associated with the Mid-Atlantic Ridge include Iceland, the Azores, the Canary and Cape Verde Islands off West Africa, Ascension Island, and Tristan de Cunha.

The volcanic islands in the middle of the North Atlantic that comprise the Azores were created by a mantle plume or hot spot that once lay beneath Newfoundland, which then drifted westward as the ocean floor spread apart at the Mid-Atlantic Ridge. The Sts. Peter and Paul islands in the mid-Atlantic north of the equator are not volcanic in origin but instead are fragments of the upper mantle uplifted near the intersection of the St. Paul transform fault and the Mid-Atlantic Ridge.

Iceland is a broad volcanic plateau of the Mid-Atlantic Ridge that rose above the sea about 16 million years ago when the ridge assumed its present position (Fig. 5–13). The island is unique because it straddles a spreading ridge system, where the two plates of the Atlantic Basin and adjacent continents pull apart. Along the ridge, the abnormally elevated topography extends in either direction about 900 miles, and over a third of of the plateau lies above sea level. South of Iceland, the broad plateau tapers off to form a structure more typical of the Mid-Atlantic Ridge.

A steep-sided, V-shaped valley runs northward across the entire length of the Iceland, and is one of the few expressions of a midocean rift on land.

Numerous volcanoes flank the rift, making Iceland one of the most volcanically active places on Earth. On other parts of the midocean ridge, volcanic activity is quite prevalent, with perhaps as many as 20 major deep underwater eruptions a year.

Volcanoes formed on or near the midocean ridges often develop into isolated peaks as they move outward from the ridge axis during seafloor spreading. The ocean floor thickens as it moves away from the spreading ridge axis. This thickening of the seafloor influences a volcano's height, because a thicker crust can support a greater mass. The ocean crust also bends like a rubber mat under the massive weight of a seamount. For instance, the crust beneath Hawaii bulges in a downward concave shape as much as 6 miles.

A volcano formed at a midocean ridge cannot increase its mass unless it continues to be supplied with magma after it leaves the vicinity of the ridge. Sometimes a volcano formed on or near a midocean ridge develops into an

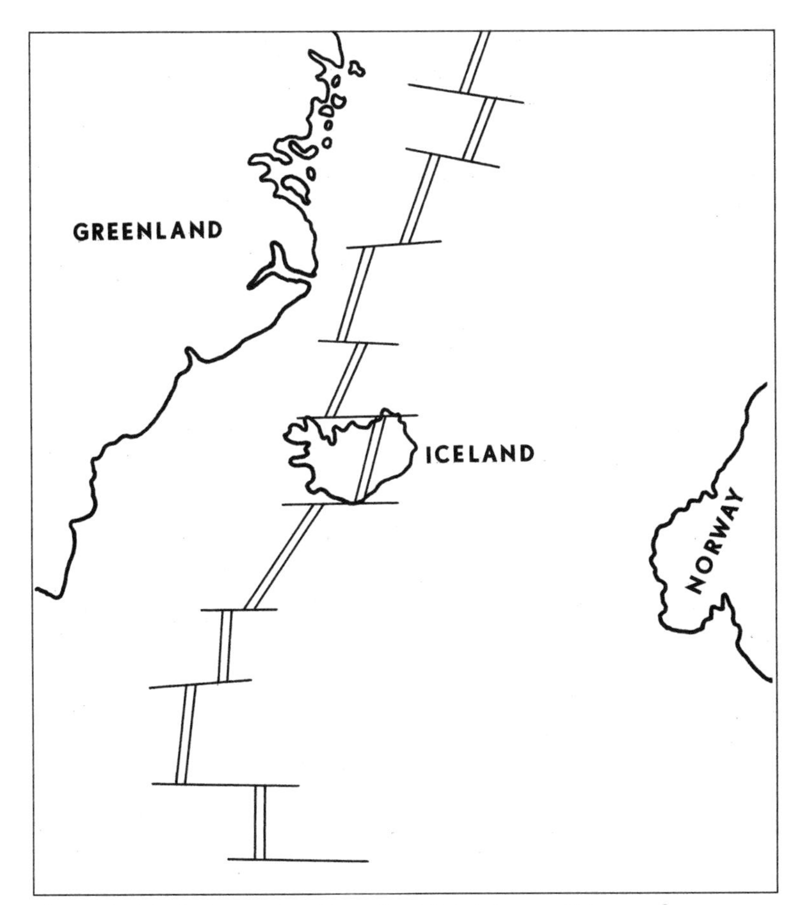

Figure 5–13 Iceland straddles the Mid-Atlantic Ridge.

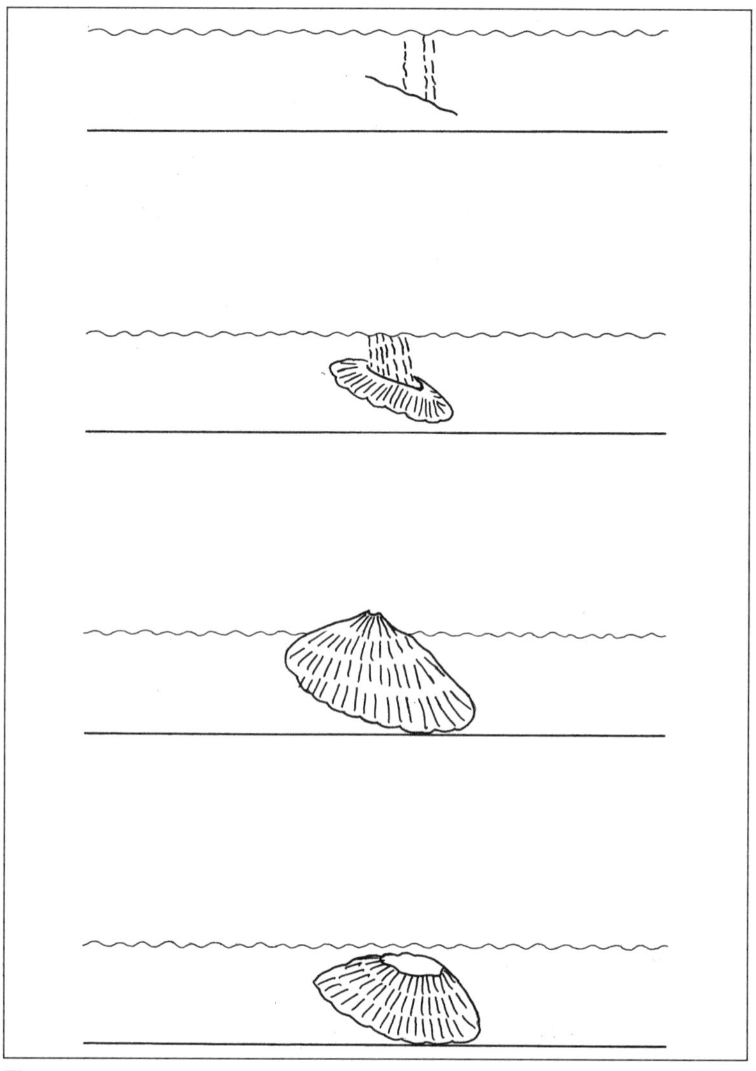

Figure 5–14 The life cycle of an oceanic volcano. From the top: A rift first forms on the ocean floor, lava piles up until the volcano rises above sea level, and the dormant cone sinks below sea level.

island, only to have its source of magma cut off. Then erosion begins to wear it down until it finally sinks beneath the sea (Fig. 5–14).

HOT SPOT VOLCANOES

About 100 small regions of isolated volcanic activity known as hot spot volcanoes exist in various parts of the world. The hot spots provide a pipeline for transporting heat from the planet's core to the surface. The plumes do not rise through the mantle as a continuous stream but, rather, as separate giant bubbles of hot rock. When the bubbles reach the ocean floor at the top of the mantle they create a secession of oceanic islands.

The ascending mantle plumes can lift an entire region. This has happened in a 3,000-mile-wide section of the South Pacific floor where several hot spots have erupted to form the Polynesian island chains. Similar swells occur under the Hawaiian chain in the North Pacific, Iceland in the North Atlantic, and the Kerguelen Islands in the southern Indian Ocean. The most active modern hot spots lie beneath the big island of Hawaii and Reunion Island to the east of Madagascar.

Unlike most other active volcanoes, those created by hot spots are rarely sited at plate boundaries but reside deep in the interiors of lithospheric plates (Fig. 5–15). Hot spot volcanoes are notable for their very geological isolation far from normal centers of volcanic and earthquake activity. Lavas of hot spot volcanoes differ markedly from those of subduction zones and rifts. The distinctive composition of hot spot magmas suggests that their source is outside the general circulation of the mantle.

The lavas comprise basalts that contain larger amounts of alkali minerals such as sodium and potassium, indicating that their source material is not connected with plate margins. Instead, the hot spots derive their source material from deep within the mantle, possibly near the top of the core. Hot spot plumes also might arise from stagnant regions in the center of convection cells or from below the region in the mantle stirred by convection currents.

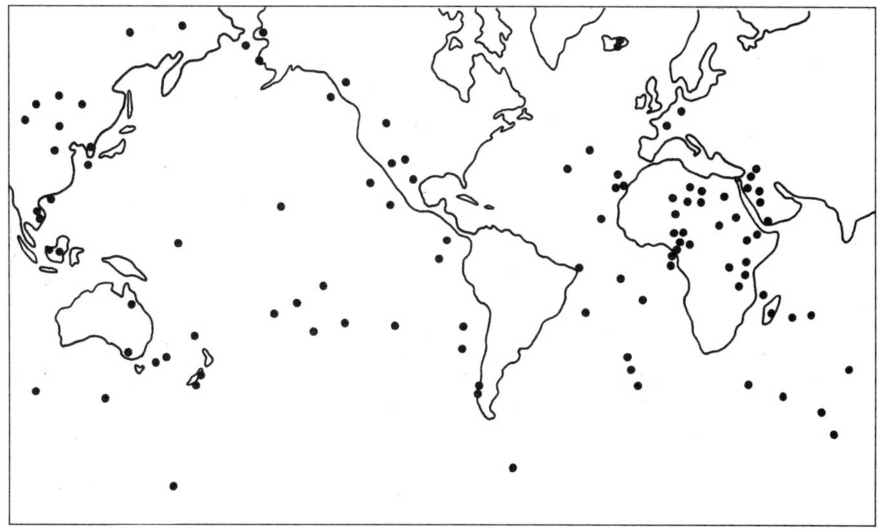

Figure 5–15 The world's hot spots, where mantle plumes rise to the surface.

As plumes of mantle material flow upward into the asthenosphere, the portion rich in volatiles rises toward the surface to feed hot spot volcanoes. The plumes exist in a range of sizes that might indicate the depth of their source material. They are not necessarily continuous flows of mantle material but might consist of molten rock, rising in giant blobs or diapirs. If the upwelling plumes stopped feeding the asthenosphere with a continuing flow of mantle material, the plates would grind to a complete halt.

The typical lifespan of a plume is a few hundred million years. Sometimes a hot spot fades away and a new one forms in its place. The position of a hot spot changes slightly as it sways in the convective currents of the

Figure 5–16 Photograph of the Hawaiian Island chain looking south, taken from the space shuttle. The main island Hawaii is in the upper portion of the photo. Courtesy of NASA

mantle. As a result, the hot spot tracks on the surface might not always be linear. However, compared with the motion of the plates, the mantle plumes are virtually stationary. Because the motion of the hot spots is so slight, they provide a reference point for determining the direction and rate of plate travel.

The passage of a plate over a hot spot often results in a trail of volcanic features, whose linear trend reveals the direction of plate motion. This produces volcanic structures aligned in a direction that is oblique to the adjacent midocean ridge system rather than parallel to it like rift volcanoes. The hot spot track might be a continuous volcanic ridge or a chain of volcanic islands and seamounts that rise high above the surrounding seafloor. The hot spot track also can weaken the crust, cutting through the lithosphere like a geologic blowtorch.

The most prominent and easily recognizable hot spot created the Hawaiian Islands, the largest islands of their kind in the world (Fig. 5–16). In effect, Hawaii's Mauna Kea Volcano, which built most of the main island, is the world's tallest mountain. It rises 6 miles above the ocean floor, exceeding the height of Mount Everest by more than half a mile.

The youngest and most volcanically active island is Hawaii, at the southeast end of the chain. Some 20 miles south of Hawaii lies a submerged volcano called Loihi, which rises about 8,000 feet above the ocean floor but is still 3,000 feet below the sea surface. Perhaps in another 50,000 years it will rise above the sea and take its place in the Hawaiian chain. The rest of the Hawaiian islands are progressively older, with extinct volcanoes trailing off to the northwest.

The entire Hawaiian chain apparently formed from a single magma source, over which the Pacific plate has passed in a northwesterly direction. The volcanic islands slowly popped out on the ocean floor conveyor belt fashion, with the oldest trailing off to the northwest, and now farthest away from the hot spot. Similar chains of volcanic islands lie in the Pacific and trend in the same general southeast-to-northwest direction as the Hawaiian Islands (Fig. 5–17), indicating that the Pacific plate is moving off in this direction. Lying parallel to the Hawaiian chain are the Austral and Tuamotu ridges. The islands and seamounts were formed by the northwestward motion of the Pacific plate over a volcanic hot spot.

The plate did not always travel in this direction, however. More than 40 million years ago it followed a more northerly heading. The course change might have resulted from a collision between the Indian and Asian plates, and appears as a distinct bend in the hot spot tracks. A sharp bend in the long Mendocino Fracture Zone jutting out from northern California confirms that the Pacific plate abruptly changed direction at the same time as the India-Asia plate convergence. The timing is also coincident with the collision of the North American and Pacific plates. From these observa-

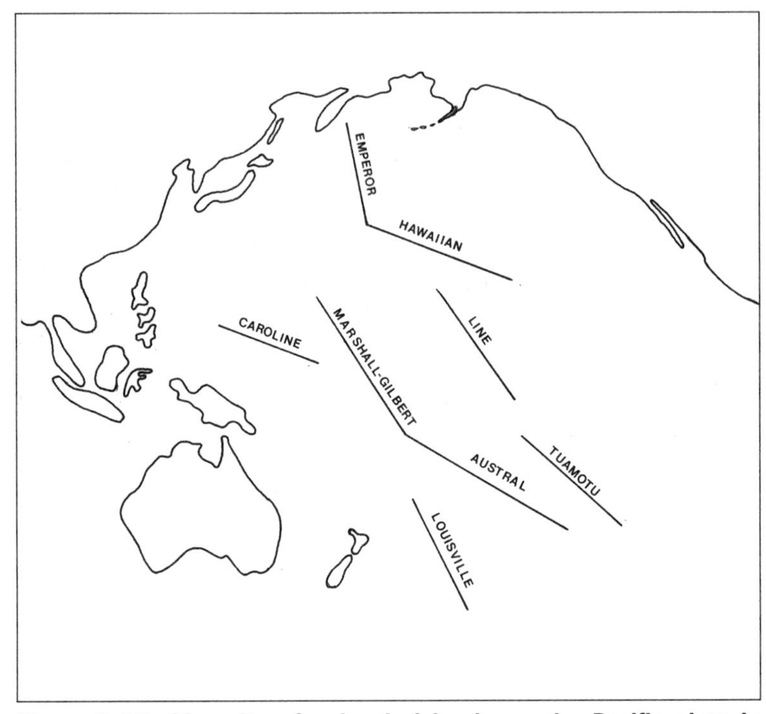

Figure 5–17 Linearity of volcanic islands on the Pacific plate in direction of movement.

tions, geologists conclude that hot spots are generally a reliable means for determining plate activity.

The Bermuda Rise in the western Atlantic, however, appears to be a contradiction to this rule. Oriented in a roughly northeast direction, parallel to the continental margin off the eastern United States, the Bermuda Rise is nearly 1,000 miles long and rises some 3,000 feet above the surrounding seafloor. The last of its volcanoes ceased erupting about 25 million years ago. A weak hot spot unable to burn a hole through the North American plate apparently was forced to take advantage of previous structures on the ocean floor acting as conduits, which explains why the volcanoes trend almost at right angles to the motion of the plate.

The Bowie seamount is the youngest in a line of submerged volcanoes running toward the northwest off the west coast of Canada. It is fed by a mantle plume more than 400 miles below the ocean floor and nearly 100 miles in diameter. But rather than lying directly beneath the seamount, as plumes usually do, this plume lies about 100 miles east of the volcano. It is believed that the plume might have taken a tilted path upward, or that the seamount somehow moved with respect to the hot spot's position.

If a midocean ridge passes over a hot spot, the plume augments the flow of molten rock welling up from the asthenosphere to form new crust. The crust is therefore thicker over the hot spot than it is along the rest of the ridge, resulting in a plateau rising above the surrounding seafloor. The Ninety East Ridge, named for its location at 90 degrees east longitude, is a succession of volcanic outcrops that runs 3,000 miles south of the Bay of Bengal and formed when the Indian plate passed over a hot spot on its way to Asia about 120 million years ago, creating an immense lava field on India known as the Rajmahal Traps.

The movement of the continents was more rapid than it is today, with perhaps the most vigorous plate tectonics the world has ever known. About 120 million years ago, an extraordinary burst of submarine volcanism struck the Pacific Basin, releasing vast amounts of gas-laden lava onto the ocean floor. The volcanic spasm is evidenced by a collection of massive undersea lava plateaus that formed almost simultaneously. The largest of

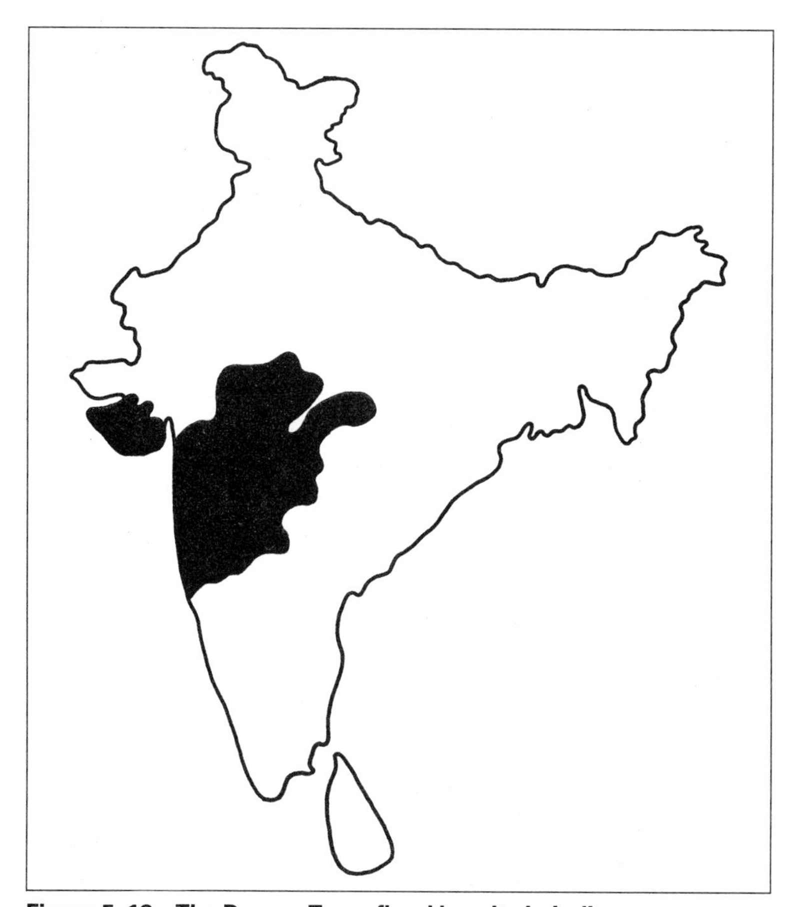

Figure 5–18 The Deccan Traps flood basalts in India.

these plateaus is the Ontong Java, northeast of Australia. Roughly two-thirds the size of the continent, it contains at least 9 million cubic miles of basalt, enough to bury the entire United States under 15 feet of lava.

About 65 million years ago, a giant rift opened up along the western side of India, and huge volumes of molten lava poured onto the surface, forming the Deccan Traps Flood basalts (Fig. 5–18). The rift separated the Seychelles Bank from the mainland, creating the Seychelles Islands. They were followed 40 million years ago by the Kerguelen Islands as India continued to trek northward toward southern Asia.

The Kerguelen plateau is the world's largest submerged platform. Approximately 50 million years ago, a huge submerged plateau in the Indian Ocean separated into two platforms that now sit about 1,200 miles apart. The plateau grew from the ocean floor more than 90 million years ago, when a series of volcanic eruptions poured out voluminous amounts of molten basalt onto the Antarctic plate as the continent separated from Australia.

During the next several million years, a long rift sliced through the plate and cut off its northern section, which latched onto the Indian plate and started on a long journey northward. Meanwhile, the southern half of the plate continued to move southward. Half of the original platform, called the Broken Ridge, currently lies off the west coast of Australia. The other half, the Kerguelen plateau, sits north of Antarctica. The Exmouth plateau is a submerged feature that sits on a sunken piece of the Australian continent, which itself was attached to India when all continents were assembled into Pangaea.

6

ABYSSAL CURRENTS

The ocean is continuously in motion, distributing water and heat to all corners of the globe. In effect, the ocean acts as a huge circulating machine that makes the Earth's climate equitable. Ocean currents follow well defined courses, transporting tremendous quantities of seawater, serving as a global "conveyor belt" over the planet.

Abyssal storms stir the deep ocean floor, shifting sediments on the seabed. El Niño currents, caused by a great sloshing of seawater in the Pacific Basin, generate unusual atmospheric weather patterns throughout the world. Waves and tides are constantly changing and rearranging the shoreline. Tsunamis produced by undersea earthquakes and coastal volcanic eruptions are among the most damaging waves, inflicting death and destruction to many seacoast inhabitants.

RIVERS IN THE ABYSS

Currents in the upper regions of the ocean (Fig. 6–1) are driven by the winds, which impart their momentum to the ocean's surface. The currents do not flow in the wind direction but are deflected by the Coriolis effect to the right of the wind direction, or to the northwest, in the Northern

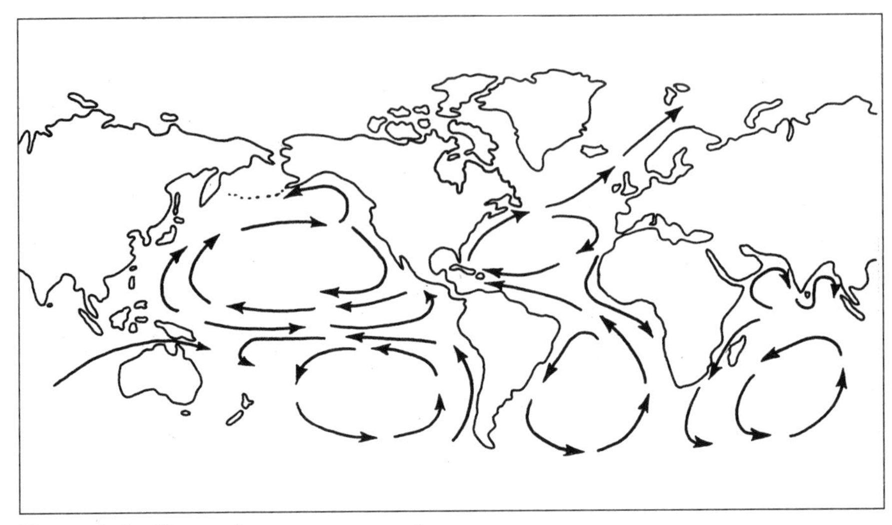

Figure 6–1 The major ocean currents.

Hemisphere and to the left of the wind direction, or to the southwest, in the Southern Hemisphere. The currents acquire warm water from the tropics, distribute it to the higher latitudes, and return to the tropics with cold water. This exchange moderates the temperatures of coastal regions, making countries like Japan and northern Europe warmer than they otherwise would be for their latitudes.

The Gulf Stream snakes 13,000 miles clockwise around the North Atlantic Basin, transporting warm tropical water to the northern regions. Its counterpart in the North Pacific is another strong current called the Japan current. This current bears warm water from the tropics, sweeps northward against Japan, crosses the upper Pacific, and turns southward to warm the western coast of North America. The major current in the South Pacific is the Humboldt or Peru current, which flows northward along the west coast of South America.

Like huge undersea tornadoes, eddies or gyres of swirling warm and cold water accompany the ocean currents. Many eddies are enormous—as much as 100 miles or more across and reaching depths of 3 miles. Most eddies, however, are less than 50 miles across; some, including those in the Arctic Ocean off Alaska, are only 10 miles wide. These small eddies play an important role in mixing the oceans like giant eggbeaters.

The eddies appear to be pinched-off sections of the main ocean currents. Like high pressure systems in the atmosphere, the eddies rotate clockwise in the Northern Hemisphere and counterclockwise in the Southern Hemisphere. Sea life caught in the eddies is often transported to hostile environments and can only survive as long as the eddies with their more favorable waters continue to operate, perhaps upwards of several months.

The world's ocean is filled nearly to the top with icy water only a few degrees above freezing that was chilled while at the surface of the polar seas. The sinking of cold, dense water near the poles generates strong, deep currents that flow steadily toward the equator (Fig. 6–2). Associated with these currents are eddies on the western side of ocean basins that are often over 100 times stronger than the main current.

In the polar regions, the surface water is denser than in other parts of the world because of its low temperature and high salt content. The increased saltiness results from the evaporation of poleward flowing water and the exclusion of salt from ice as it freezes. As seawater increases in density, it sinks to the bottom, then spreads out upon hitting the ocean floor, and heads toward the equator. The Coriolis effect deflects global currents westward because of the Earth's eastward rotation. The distribution of landmasses and the topography of the ocean floor, including ridges and canyons, also affect the path taken by the circulating water.

The Antarctic plays a larger role in global ocean circulation than does the Arctic. Deep cold currents flowing from Antarctica toward the equator trend to the left and press against the western side of the Atlantic, Pacific, and Indian ocean basins. As they sweep against the continents, the currents pick up speed similar to the way in which a stream flows faster in a narrowing channel.

Swift-flowing currents along some parts of the ocean bottom are still much of a mystery. One deep current, after traveling 7,500 miles from its source in the Antarctic, turns and sweeps along the edge of the abyssal plain

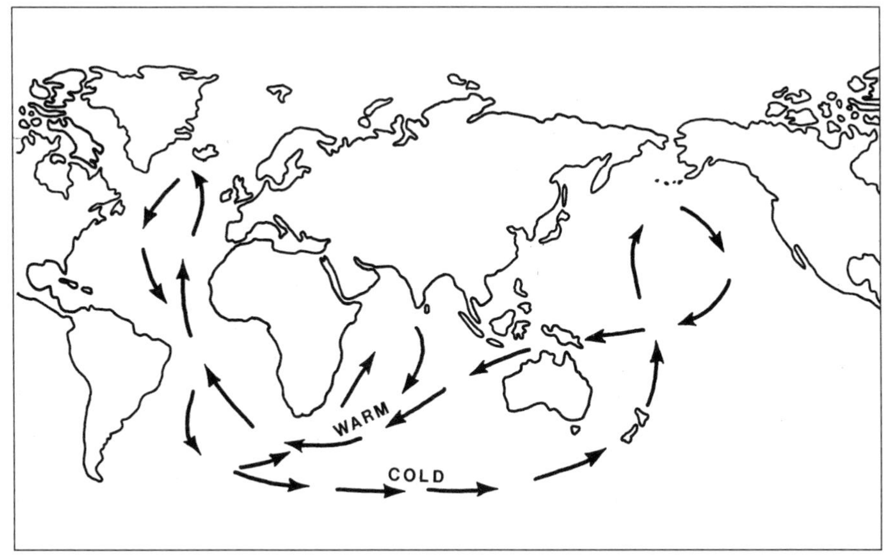

Figure 6–2 The ocean conveyor belt, transporting warm and cold water over the Earth.

south of Nova Scotia. The Atlantic Bottom Water, the largest mass of bottom water in the world, sinks from the surface near Antarctica and flows northward along the seafloor into the western North Atlantic.

Before mixing with North Atlantic water and dispersing, some of this flow curves to the west (due to the Earth's eastward rotation) and hugs the lower edge of the continental rise at the border of the abyssal plain. This current, along with the lower reaches of intense eddies pinched off the Gulf Stream, might account for the muddy waters kicked up from the nearly 2,000-foot-deep abyss south of Nova Scotia that extends as far south as the Bahamas. Deep eddies induced by the Gulf Stream might be superimposed on this flow to produce undersea storms of sediment-laden water.

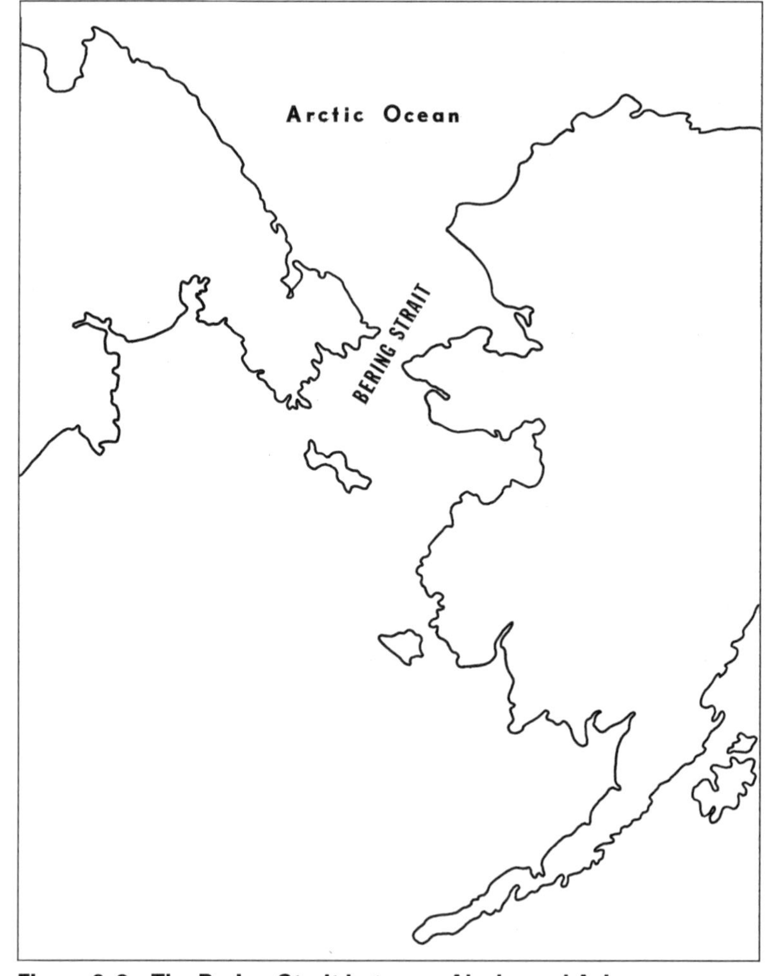

Figure 6–3 The Bering Strait between Alaska and Asia.

The Indian Ocean is unique because it is not in contact with the north polar region and has only one source of cold bottom water from the Antarctic. By contrast, the Atlantic and Pacific connect with both the Antarctic and the Arctic oceans. The narrow, shallow Bering Strait that separates Alaska from Asia (Fig. 6–3) blocks the flow of deep cold water from the Arctic Ocean into the Pacific. Seawater freezes more readily in the Arctic regions because the near-surface water is not sufficiently enriched in salt and therefore not dense enough to sink. Consequently the Arctic Ocean is largely a sea of ice.

Seawater in the Atlantic is saltier than in the Pacific because of the greater contribution of river borne salts. The Atlantic has two major sources of highly saline water. One is the Gulf of Mexico, whose water is carried northward by the Gulf Stream, and the other is the deep flow from the Mediterranean Sea. The climate of the Mediterranean is so warm that evaporation concentrates salt in that sea, and Mediterranean water spilling westward through the Strait of Gibraltar sinks to a depth of nearly 4,000 feet in the Atlantic. These sources raise the salt content of the surface waters of the North Atlantic to levels much higher than in the North Pacific.

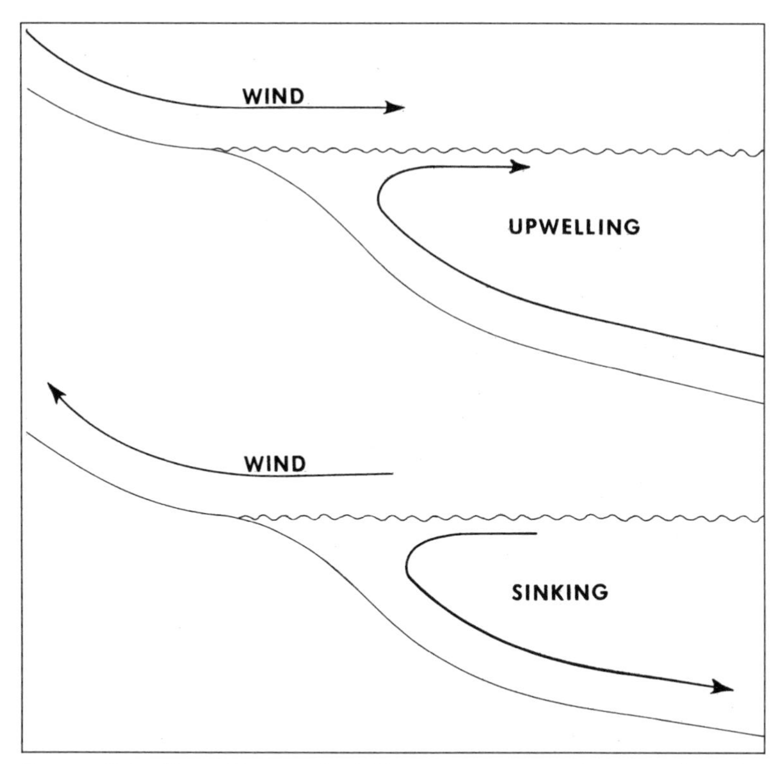

Figure 6–4 Upwelling and sinking ocean currents are driven by offshore and onshore winds.

As the surface water of the North Atlantic moves northward, it enters the Norwegian Sea, where it cools below the freezing point of fresh water but, because of its high salt content, does not freeze. The cold, dense water sinks, and upon reaching the bottom it reverses direction and flows back into the Atlantic through a series of deep, narrow troughs in the submarine ridges that connect Greenland, Iceland, and Scotland. This deep-sea current, called the North Atlantic Deep Water, is a subsurface stream with a flow 20 times greater than all the world's rivers combined.

As this large volume of deep water moves southward, it flows to the right against the continental margin of eastern North America, forming the Western Boundary Undercurrent. This current transports some 20,000 cubic miles of water annually along the east coast of North America. All these deep-ocean currents travel very slowly, completing the journey from the poles to the equator and back again in upward of 1,000 years, as compared to surface currents, which complete the circuit around an ocean basin in less than a decade.

The volume of rising water in parts of the ocean matches the volume of sinking water in the polar regions. The cold waters from the polar seas rise in upwelling zones in the tropics, creating an efficient heat transport system. Upwelling currents off the coasts of continents and near the equator are important sources of bottom nutrients. Modern fishermen track down these areas of upwelling water, which is usually where the fish are.

The tropical seas are warmed by solar radiation from above and cooled by upwelling water from below. This interaction gives rise to an equator-to-pole cycle of heat transport. Offshore and onshore winds also drive upwelling currents (Fig. 6–4). These processes involve the entire ocean in a gigantic heat engine, transporting a tremendous amount of heat around the globe.

EL NIÑO

Ocean currents dramatically affect the climate, and major changes in these systems can cause abnormal weather patterns all over the world (Fig. 6–5). Unusual oceanographic conditions during the 1982–83 El Niño dramatically affected the Galapagos Islands in the Pacific, off the coast of Ecuador. The ocean current patterns around the islands are complex and are greatly influenced by the equatorial undercurrent, a subsurface, eastward-flowing current about 600 feet thick. During a period when the sea surface temperatures were anomalously high, a major redistribution of phytoplankton around the Galapagos Islands might have contributed to the reproductive failure of seabirds and marine mammals on the islands.

About every 3 to 7 years, anomalous atmospheric pressure changes known as El Niño Southern Oscillation (ENSO) occur in the South Pacific. As atmospheric pressure rises on Easter Island in the eastern Pacific, it falls

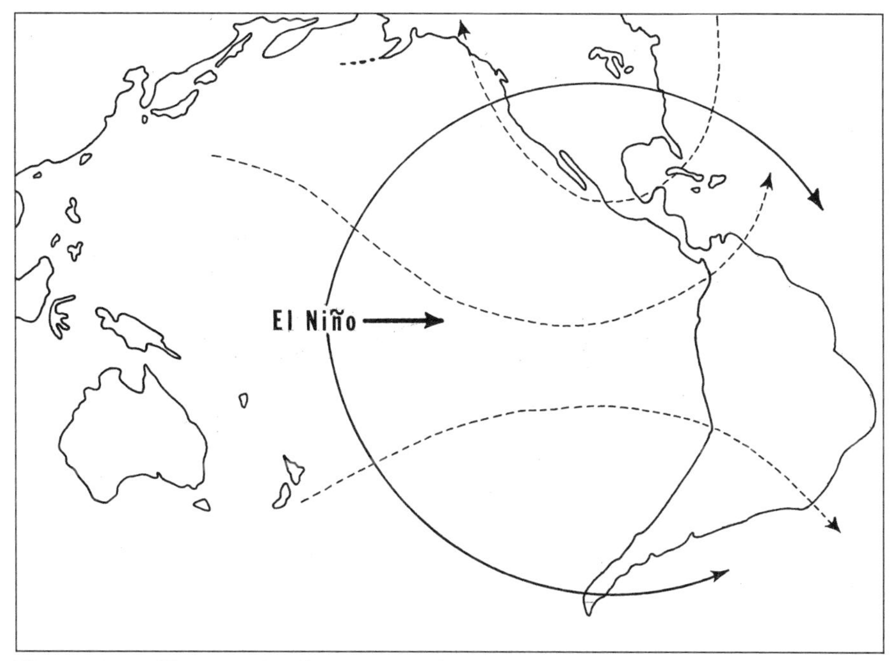

Figure 6–5 Changes in air currents during El Niño. Dashed lines are normal currents.

in Darwin, Australia in the western Pacific. When a major El Niño occurs, the barometric pressure over the eastern Pacific falls, while the pressure over the western Pacific rises. When El Niño ends, the pressure difference between these two areas reverses, creating a massive seesaw effect of atmospheric pressure. While the El Niño is in process, it disrupts the westward-flowing trade winds. Warm water piled up in the western Pacific by the wind force flows back to the east, creating a great sloshing of water in the South Pacific Basin.

The opposite condition occurs during a La Niña, when the surface waters of the Pacific cool. In mid-1988, water temperatures in the central Pacific plummeted to abnormally low levels, signaling a climate swing from an El Niño to a La Niña, and the changing climate affected the precipitation-evaporation balance of the world (Fig. 6–6). Strong monsoons hit India and Bangladesh, and heavy rains visited Australia. The La Niña might also have been responsible for a severe drought in the United States and a marked drop in global temperatures a year later.

Along the west coast of South America, the southeast trade winds drive the Peru Current, pushing surface water offshore and allowing cold, nutrient-rich water to well up to the surface. The westward push of the trade winds continues across the eastern and central Pacific, and the resulting stress on the sea surface piles up water in the western Pacific, causing the

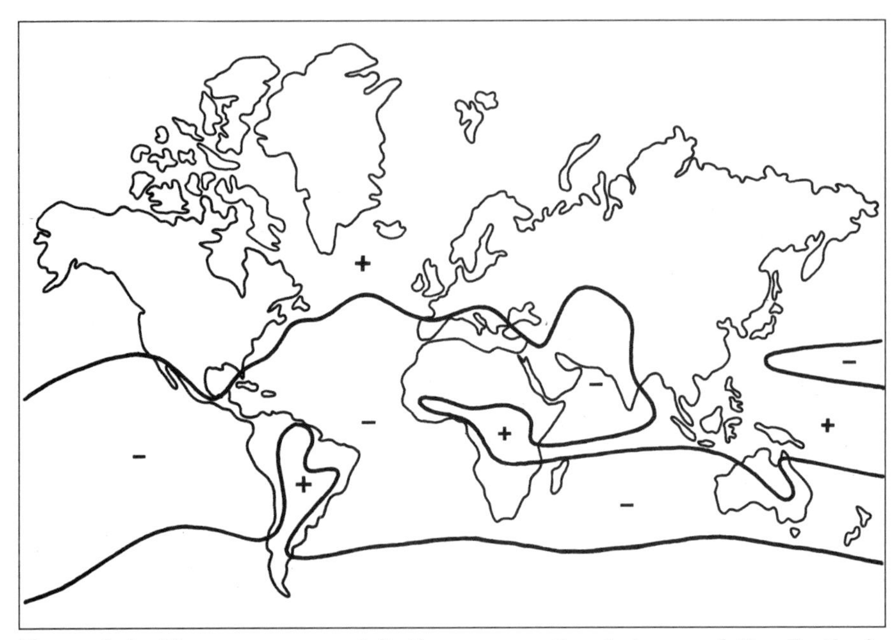

Figure 6–6 The average precipitation-evaporation balance of the Earth. In positive areas, precipitation exceeds evaporation. In negative areas, evaporation exceeds precipitation.

warm surface layer of the ocean to thicken in the west and thin in the east. The thermocline, the boundary between the cold and warm layers of the ocean, falls to about 600 feet in the western Pacific and rises to about 150 feet in the eastern Pacific. Because the thermocline is near the surface, the upwelling waters off the coast of South America are usually cold.

During an El Niño event beginning around October, the cessation of the trade winds in the western Pacific causes the thick layer of warm water to collapse. The water flows back toward the east in subsurface waves called Kelvin waves that reach the coast of South America in 2 to 3 months, creating a huge reverse flow of seawater in the South Pacific Basin. The Kelvin waves generate eastward-flowing currents that transport warm water from the west. This lowers the level of the thermocline and prevents the upwelling of cool water from below.

With an increase of warm water from the west and the suppression of cold water from below, the sea surface begins to warm considerably by December or January. The stretch of warm water shifts the position of thunderstorms that pump heat and water into the atmosphere, thus rerouting atmospheric currents around the world. As the El Niño continues to develop, the trade winds begin blowing from the west, intensifying the Kelvin waves and further depressing the thermocline off South America.

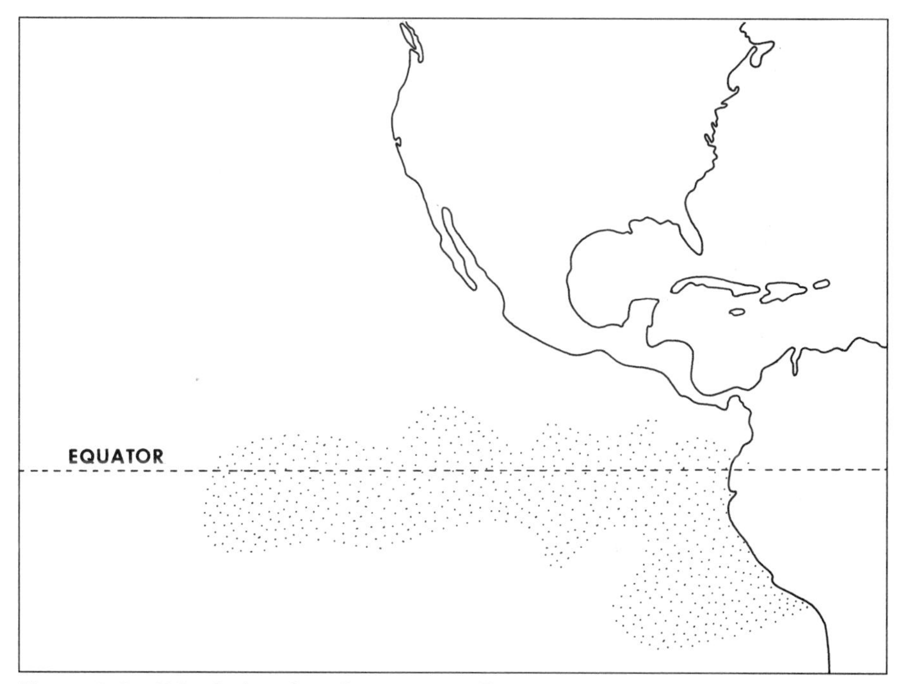

Figure 6–7 Stippled region shows area affected by increased sea-surface temperature during the 1972 El Niño in the Pacific Ocean.

The Peru current, flowing northward along the west coast of South America, is not significantly weakened by the El Niño and continues to pump water to the surface, though this time the upwelling water is warm, causing a major decline of fisheries. The westward current off equatorial South America is not only weakened by the eastward push of the Kelvin waves but is also much warmer than before. This spreads the warming of the sea surface westward along the equator (Fig. 6–7), and the normal wind pattern reverses, causing a major disruption in global weather patterns.

ABYSSAL STORMS

The dark abyss at the bottom of the ocean was thought to be quiet and almost totally at rest, with sediments slowly raining down and accumulating at a rate of about 1 inch in 20 centuries. Recent discoveries reveal signs that infrequent undersea storms often shift and rearrange the sedimentary material that has rested on the bottom for long periods. Occasionally, the surging bottom currents scoop up the top layer of mud, erasing animal tracks and creating ripple marks in the sediments, much like those produced by wind and river currents.

On the western side of the ocean basins, undersea storms skirt the foot of the continental rise, transporting huge loads of sediment and dramatically modifying the seafloor. The storms scour the ocean bottom in some areas and deposit large volumes of silt and clay in others. The energetic currents travel at about 1 mile per hour; because of the high density of seawater, they sweep the ocean floor just as effectively as a gale with winds up to 45 miles per hour erodes shallow areas near shore.

That abyssal storms seem to follow certain well-traveled paths is indicated by long furrows of sediment on the ocean floor (Fig. 6–8). The scouring of the seabed and deposition of thick layers of fine sediment results in much more complex marine geology than that developed simply from a constant rain of sediments. The periodic transport of sediment creates layered sequences that look similar to those created by strong windstorms in shallow seas, with overlapping beds of sediment graded into different grain sizes.

Sedimentary material deposited on the ocean floor consists of detritus, which is terrestrial sediment and decaying vegetation, along with shells and skeletons of dead microscopic organisms that flourish in the sunlit waters of the top 300 feet of the ocean. The ocean depth influences the rate of marine-life sedimentation. The farther the shells descend, the greater their chance of dissolving in the cold, high-pressure waters of the abyss before reaching the bottom. Preservation also depends on rapid burial and protection from the corrosive action of the deep-sea water.

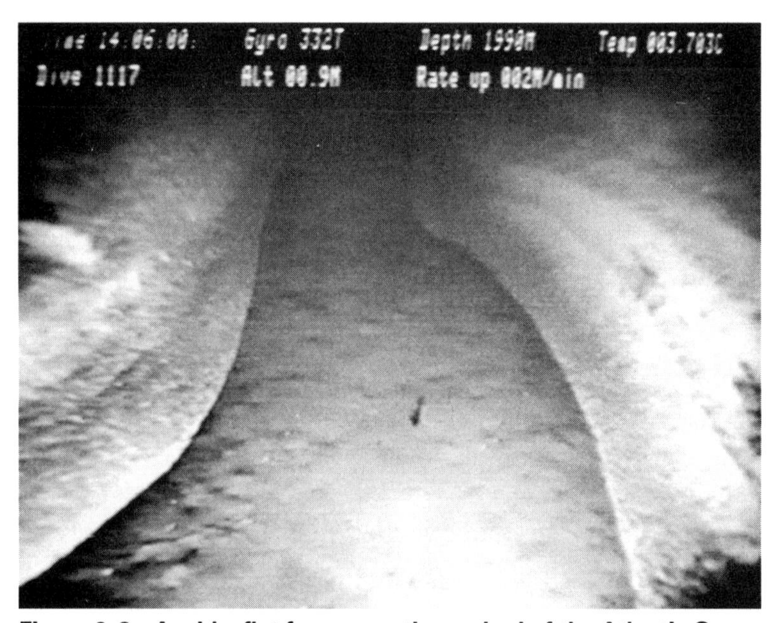

Figure 6–8 A wide, flat furrow on the seabed of the Atlantic Ocean from the deep submersible *Alvin*. Photo by N. P. Edgar, courtesy of USGS

Figure 6–9 The Yahtse River delta, Icy Bay, Alaska. Photo by J. H. Hartshorn, courtesy of USGS

Rivers carry detritus to the edge of the continent and out onto the continental shelf, where marine currents pick up the material. When the detritus reaches the edge of the shelf, it falls to the base of the continental rise under the pull of gravity. Approximately 15 billion tons of continental material reaches the mouths of rivers and streams annually (Fig. 6–9). Most of this detritus is deposited near the river outlets and on continental shelves; only a few billion tons falls into the deep sea. In addition to the river-borne sediments, strong desert winds in subtropical regions sweep out to sea a significant amount of terrestrial material. The windblown sediment also contains significant amounts of iron, an important nutrient that supports prolific blooms of plankton. In iron-deficient parts of the ocean, there are "deserts" where "jungles" should have been even though plenty of other nutrients exist.

The biological material in the sea contributes about 3 billion tons of sediment to the ocean floor each year. The biologic productivity, controlled in large part by the ocean currents, governs the rates of accumulation. Nutrient-rich water upwells from the ocean depths to the sunlit zone, where microorganisms ingest the nutrients. Areas of high productivity and high rates of accumulation normally occur near major oceanic fronts, such as the region around Antarctica, and along the edges of major currents, such as the Gulf Stream and the Kuroshio or Japan current that circles clockwise around the North Pacific Basin.

The greatest volume of silt and mud and the strongest bottom currents are in the high latitudes of the western side of the North and South Atlantic. These areas have the largest potential for generating abyssal storms that

Figure 6–10 An instrument that measures water dynamics and sediment mobilization on the ocean floor. Photo by N. P. Edgar, courtesy of USGS

form and shape the seafloor. They also have the biggest drifts of sediment on Earth, covering an area more than 600 miles long, 100 miles wide, and 1 mile thick. Abyssal currents at depths of 2 to 3 miles play a major role in shaping the entire continental rise off North and South America. Elsewhere in the world, bottom currents shape the distribution of fine-grained material along the edges of Africa, Antarctica, Australia, New Zealand, and India.

Instruments lowered to the ocean floor measure water dynamics and their effects on sediment mobilization (Fig. 6–10). During abyssal storms, the velocity of bottom currents increases from about one-tenth to over 1 mile per hour. The storms in the Atlantic seem to derive their energy from eddies that emerge from the Gulf Stream. While the storm is in progress, the suspended sediment load increases tenfold, and the current is able to carry about 1 ton of sediment per minute for long distances. The moving clouds of suspended sediment appear as coherent patches of turbid water with a residence time of about 20 minutes. The storm itself might last from several days to a few weeks, at the end of which the current velocity slows to normal and the sediment drops out of suspension.

Not all drifts are directly attributable to abyssal storms. Material carried by deep currents have modified vast areas of the ocean as well. The storm's main effect is to stir sediment that bottom currents then pick up and carry downstream for long distances. The circulation of the deep ocean does not show a strong seasonal pattern; therefore, the onset of abyssal storms is unpredictable. Abyssal storms are likely to strike an area every 2 to 3 months.

TIDAL CURRENTS

Tides result from the pull of gravity of the moon and sun on the ocean. The moon revolves around the Earth in an elliptical orbit and exerts a stronger pull on the side facing the moon than on the opposite side. The difference between the moon's gravitational attraction on both sides is about 13 percent, which elongates the center of gravity of the Earth-moon system. The pull of gravity creates two tidal bulges on the Earth. As the Earth revolves, the oceans flow into the two tidal bulges, one facing toward the moon and the other facing away from it. Between the tidal bulges, the ocean is shallower, giving it an overall egg-shaped appearance. The middle of the ocean only rises about 2.5 feet at maximum high tide; but due to a sloshing-over effect and the configuration of the coastline, the tides on the coasts often rise several times higher.

The daily rotation of the Earth causes every point on the surface to go into and out of the two tidal bulges once every day. Thus, as the Earth spins into and out of each tidal bulge, the tides appear to rise and fall twice daily. The moon also orbits the Earth in the same direction that it rotates, only faster. By the time a point on surface has rotated halfway around, the tidal bulges have moved forward with the moon, and the point must travel farther each day to catch up with the bulge. Therefore, the actual period between high tides is 12 hours 25 minutes.

If continents did not impede the motion of the tides, all coasts would have two high tides and two low tides of nearly equal magnitudes and durations each day. These are called semidiurnal tides, and occur at places such as along the Atlantic coasts of North America and Europe. However, different tidal patterns form when the tide wave is deflected and broken up by the continents. Because of this action, the tidal ocean forms a complicated series of crests and troughs thousands of miles apart. In some regions, the tides are coupled with the motion of large nearby bodies of water. As a result, some areas—for example, the coast of the Gulf of Mexico—have only one tide a day, called a diurnal tide, with a period of 24 hours 50 minutes.

The sun also raises tides with semidiurnal and diurnal periods of 12 and 24 hours. Because the sun is much farther away from the Earth, its tides are only about half the magnitude of lunar tides. The overall tidal amplitude, which is the difference between the high-water level and the low-water level, depends on the relation of the solar tide to the lunar tide and is controlled by the relative positions of the Earth, moon, and sun.

The tidal amplitude is at its maximum twice a month during the time of new and full moon, when the Earth, moon, and sun align in a nearly straight-line configuration (known as "syzygy" from the Greek word *syzygos*, meaning "yoked together"). This is the time of the spring tides—from the Saxon word *springan*, meaning "a rising or swelling of water." Neap tides occur when the amplitude is at a minimum during the first and third quarters of the moon, when the relative positions of the Earth, moon, and sun form a right angle and the solar and lunar tides oppose each other.

A tidal basin near the mouth of a river can actually resonate with the incoming tide. The oscillation makes the water at one side of the basin high at the beginning of the tidal period, low in the middle, and high again at the end of the tidal period. The incoming tide sets the water in the basin oscillating, sloshing back and forth. The motion of the tide moving in toward the mouth of the river and the motion of the oscillation are synchronized, which reinforces the tide in the bay and makes the high tides higher and the low tides lower than they would be otherwise.

Tidal bores are a special feature of this type of oscillation within a tidal basin. They are solitary waves that carry tides upstream, usually during a new or full moon. One of the largest tidal bores sweeps up the Amazon River, with waves up to 25 feet high and several miles wide reaching 500 miles upstream. Although any body of water with high tides can generate a tidal bore, only half of all tidal bores are associated with resonance in a tidal basin.

TABLE 6–1 MAJOR TIDAL BORES

Country	Tidal Basin	Tidal Body	Known Bore Location
Bangladesh	Ganges	Bay of Bengal	
Brazil	Amazon	Atlantic Ocean	
	Capim		Capim
	Canal do Norte		
	Guama		
	Tocantins		
	Araguari		
Canada	Petitcodiac	Bay of Fundy	Moncton
	Salmon		Truro
China	Tsientang	East China Sea	Haining to Hangchow
England	Severn	Bristol Channel	Framilode to
	Parret		Gloucester
	Wye		Bridgwater
	Mersey	Irish Sea	Liverpool to
	Dee		Warrington
	Trent	North Sea	Gunness to
			Gainsborough
France	Seine	English Channel	Gaudebec
	Orne		
	Coueson	Gulf of St. Malo	
	Vilaine	Bay of Biscay	
	Loire		
	Gironde		Iles de Margaux
	Dordogne		La Caune to Brunne
	Garonne		Bordeaux to Cadillac
India	Narmada	Arabian Sea	
	Hooghly	Bay of Bengal	Hooghly Pt. to Calcutta
Mexico	Colorado	California Gulf	Colorada Delta
Pakistan	Indus	Arabian Sea	
Scotland	Solway Firth	Irish Sea	
	Forth		
United States	Turnagain Arm	Cook Inlet	Anchorage to Portage
	Knik Arm		

The seaward ends of many rivers experience tides; in such cases, at the river mouth the tides are symmetrical, with ebb and flood tide lasting about 6 hours each. Ebb and flood tides refer to the currents associated with the tides. Ebb currents flow out to sea, while flood currents flow into an inlet. Upstream, the tides become increasingly asymmetrical, with less time elapsing between low water and high water than between high water and low water, as the tide comes in quickly but goes out gradually with the river current. A tidal bore exaggerates this asymmetry because the tide comes up the river very rapidly in a single wave. As the tidal bore moves upstream, it must continue to travel faster than the river current or else it will be swept downstream and out to sea.

OCEAN WAVES

Ocean waves form by large storms at sea when strong winds blow across the water's surface (Fig. 6–11). The wave fetch is distance over which the wind blows on the surface of the ocean and is dependent on the size of the

Figure 6–11 Open ocean waves and a mysterious weather phenomenon known as sea smoke, 150 miles east of Norfolk, Virginia. Courtesy of U.S. Navy

Figure 6–12 The buckled flight deck of the USS *Bennington* after a typhoon in the western Pacific in June 1945. Courtesy of U.S. Navy

storm and the width of the body of water. For waves to reach a fully developed sea state, the fetch must be at least 200 miles for a wind of 20 knots, 500 miles for a wind of 40 knots, and 800 miles for a wind of 60 knots.

The wind speed and duration determine the wave height. With a wind speed of 30 miles per hour, for example, a fully developed sea is attained in 24 hours, with wave heights up to 20 feet. The maximum sea state occurs when waves reach their maximum height, usually after 3 to 5 days of strong, steady storm winds blowing across the surface of the ocean. However, if the sustained wind blew at 60 miles per hour, a fully developed sea would have wave heights averaging over 60 feet.

The wave height, measured from the top of the crest to the bottom of the trough, is generally less than 20 feet. Occasionally, storm waves of 30 to 50 feet high have been reported, but these are not very frequent. Exceptionally large ocean waves are rare. One such wave reported in the Pacific by a U.S. Navy tanker in 1933 was over 100 feet high. Another large wave buckled the flight deck of the aircraft carrier USS *Bennington* during a typhoon in the western Pacific in 1945 (Fig. 6–12).

The wave shape varies with the water depth (Fig. 6–13). In deep water, a wave is symmetrical, with a smooth crest and trough. In shallow water, a wave is asymmetrical, with a peaked crest and a broad trough. If the water depth is more than one-half the wave length, the waves are considered

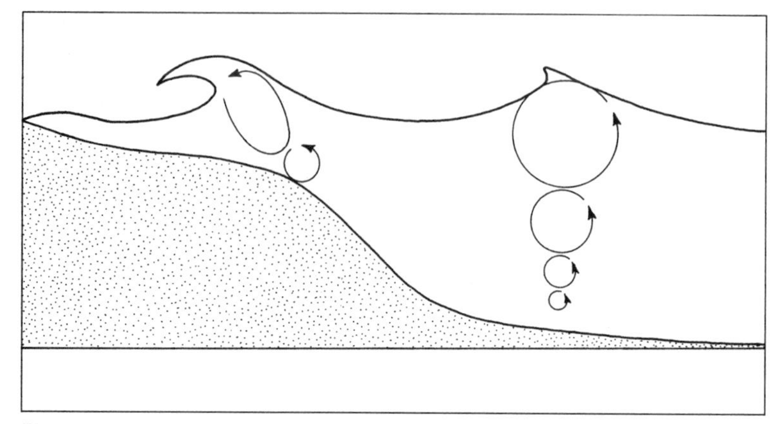

Figure 6–13 The mechanics of a breaker.

deep-water waves. If the water depth is less than one-half the wave length, the waves are called shallow-water waves.

The wave length (Fig. 6–14) is measured from crest to crest and depends on the location and intensity of the storm at sea. The average lengths of storm waves vary from 300 to 800 feet. As waves move away from a storm area, the longer waves move ahead of the storm and form swells that travel great distances. In the open ocean, swells of 1,000-foot wave-lengths are common, with a maximum of about 2,500 feet in the Atlantic and about 3,000 feet in the Pacific.

The wave period is the time a wave takes to pass a certain point and is measured from one wave crest to the next. Wave periods in the ocean vary

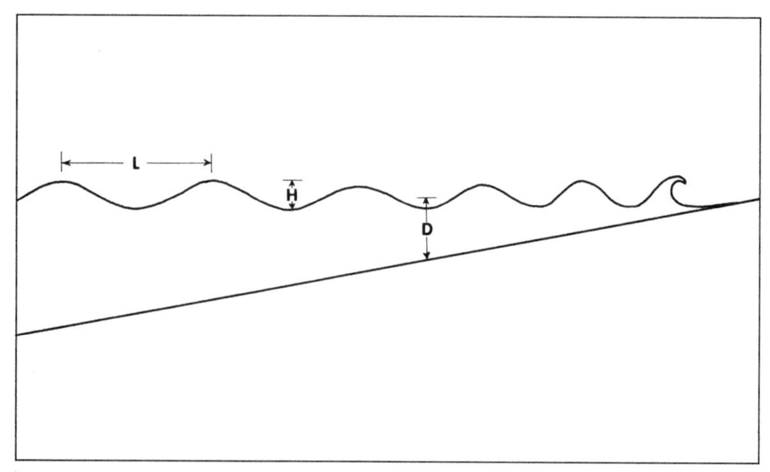

Figure 6–14 Properties of waves. L = wave length, H = wave height, D = wave depth.

from less than a second for small ripples to more than 24 hours. Waves with periods of less than 5 minutes are called gravity waves and include the wind driven waves that break against the coastline, most of which have periods between 5 and 20 seconds. A seismic sea wave from an undersea earthquake or landslide usually has a period of 15 minutes or more and a wave length of up to several hundred miles.

Waves with periods of between 5 minutes and 12 hours are called long waves, and are generated by storms. Other long waves result from seasonal differences in barometric pressure over various parts of the ocean such as the Southern Oscillation discussed in a previous section. Waves with longer periods travel faster than shorter-period waves, and the speed is proportional to the square root of the wave length. Short-period waves are relatively steep and are particularly dangerous to small boats because the bow might be on a crest while the stern is in a trough, causing it to capsize or be swamped.

SEISMIC SEA WAVES

Destructive waves also result from undersea and near-shore earthquakes (Fig. 6–15). They are called seismic sea waves or tsunamis, a Japanese word meaning "tidal waves"—so named because of their common occurrence in

Figure 6–15 Seismic sea wave damage at a railroad marshaling yard, Seward district, Alaska, from the March 27, 1964, earthquake. Courtesy of USGS

this region. The waves really have nothing to do with the tides, however. The vertical displacement of the ocean floor during earthquakes causes the most destructive tsunamis, whose wave energy is proportional to the intensity of the quake.

In the open ocean, the wave crests are up to 300 miles long and usually less than 3 feet high. However, the waves extend downward for thousands of feet, all the way to the ocean bottom. The distance between crests, or the wave length, is 60 to 120 miles, giving the tsunami a very gentle slope, which allows it to pass beneath ships practically unnoticed. Tsunamis travel at speeds of between 300 and 600 miles per hour. Upon entering

Figure 6–16 A fishing boat beached several hundred feet inland from the head of Resurrection Bay, Seward district, by seismic sea waves from the March 27, 1964, Alaskan earthquake. Courtesy of USGS

shallow coastal waters, tsunamis have been known to grow into a wall of water up to 200 feet high, although most tsunamis are only a few tens of feet high.

When a tsunami touches bottom in a harbor or narrow inlet, its speed rapidly diminishes to about 100 miles per hour. The sudden breaking action causes the water to pile up, magnifying the wave height tremendously. The destructive force of the wave is immense, and the damage it causes as it crashes into the shore is considerable. Large buildings are crushed with ease, and sizable ships are tossed up and carried well inland like toys (Fig. 6–16).

Explosive eruptions associated with the birth or the death of a volcanic island also set up large tsunamis that are highly destructive. Volcanic eruptions that develop tsunamis are responsible for about a quarter of all deaths from tsunamis. The powerful waves transmit the volcano's energy to areas outside the reach of the volcano itself. Large pyroclastic (volcanic fragment) flows into the sea or landslides triggered by volcanic eruptions produce tsunamis as well. Coastal and submarine slides also generate large tsunamis that can overrun portions of the adjacent coast (Fig. 6–17).

Figure 6–17 Wave damage shown in bare areas on Cenotaph Island and south shore of Lituya Bay, Alaska, resulting from a massive rock slide in 1958. Photo by D. J. Miller, courtesy of USGS

Large parts of Alaska's Mount St. Augustine have collapsed and fallen into the sea, generating large tsunamis. Massive landslides have ripped out the flanks of the volcano ten or more times during the past 2,000 years. The last slide occurred during the October 6, 1883, eruption, when debris from the flanks of the volcano crashed into the Cook Inlet. The impact sent a 30-foot tsunami to Port Graham 54 miles away destroying boats and flooding houses.

Until about 40 years ago, earthquakes on the ocean floor went largely undetected, and the only warning people had of a tsunami was a rapid withdrawal of water from the shore. Residents of coastal areas frequently stricken by tsunamis heed this warning and immediately head for higher ground. Several minutes after the sea retreats, a tremendous surge of water extends hundreds of feet inland. Often a succession of surges occurs, each followed by a rapid retreat of water back to the sea. On coasts and islands protected by barrier reefs or where the seafloor rises gradually, much of the tsunami's energy is spent before reaching the shore. On volcanic islands, which lie in very deep water, like the Hawaiian Islands, or where deep submarine trenches lie immediately outside harbors, an oncoming tsunami can build to prodigious heights.

The most tsunami-prone area in the world is the Pacific rim, which has the most earthquakes as well as the most volcanoes. Destructive tsunamis from submarine earthquakes can travel clear across the Pacific and reverberate through the ocean for days. A tsunami originating in Alaska could reach Hawaii in 6 hours, Japan in 9 hours, and the Philippines in 14 hours. A tsunami originating off the Chilean coast could reach Hawaii in 15 hours and Japan in 22 hours. Fortunately, this gives people in the coastal areas enough time to take the necessary safety precautions to protect life and property.

7

COASTAL GEOLOGY

The constant shifting of sediments on the surface and the accumulation of deposits on the ocean floor assures that the face of the Earth continues to change over time. Seawater lapping against the shore during a severe storm causes coastal erosion. Steep waves that accompany storms at sea erode sand dunes and sea cliffs. The continuous pounding of the surf also tears down most artificial barriers against the rising sea.

America's once sandy beaches are sinking beneath the waves. Barrier islands and sand bars running along the American East Coast and the coast of Texas are disappearing at alarming rates. Sea cliffs are eroding farther inland in California, often destroying expensive homes. Most defenses, such as seawalls erected to stop beach erosion, usually end in defeat as waves relentlessly batter the shoreline (Fig. 7–1).

SEDIMENTATION

Most sedimentary processes take place very slowly on the bottom of the ocean. The continents are mainly the sites of erosion, whereas the oceans are mostly the sites of sedimentation. Marine sediments consist of material washed off the continents, and most sedimentary rocks form along conti-

Figure 7–1 Damage to a beach area caused by storms and high tides at Virginia Beach, Virginia. Photo by K. Rice, courtesy of USDA–Soil Conservation Service

nental margins and in the basins of inland seas. Such seas invaded the interiors of North and South America, Europe, and Asia during the Mesozoic era. Areas with high sedimentation rates form deposits thousands of feet thick; where they are exposed to the surface, individual sedimentary beds can be traced for hundreds of miles.

The formation of sedimentary rock begins when erosion wears down mountain ranges and rivers carry the debris into the sea. The sediments originate from the weathering of surface rocks. The products of weathering include a wide range of materials, from very fine-grained sediments to huge boulders. Exposed rocks on the surface chemically break down into clays and carbonates and mechanically break down into silts, sands, and gravels.

Erosion by wind, rain, or glacial ice brings the sediments to streams, and the loose sediment grains travel downstream to the sea. Angular sediment grains indicate a short time spent in transit. Rounded sediment grains indicate severe abrasion from long-distance travel or from reworking by fast-flowing streams or by pounding waves on the beach. Indeed, many sandstone formations were once beach deposits.

Annually, some 25 billion tons of sediment are carried by stream runoff into the ocean and settle onto the continental shelf. The towering landform of the Himalayas is the greatest single source of sediment. Rivers draining the region—notably the Ganges and the Brahmaputra—discharge about 40 percent of the world's total amount of sediment into the Bay of Bengal, where sedimentary layers stack up miles thick.

Rivers like the Amazon and the Mississippi transport enormous quantities of sediment derived from their respective continental interiors. Large-scale deforestation and severe soil erosion at its headwaters force the Amazon of South America, the world's largest river, to carry heavier sediment loads. The Mississippi River and its tributaries drain a major section of the central United States, from the Rockies to the Appalachian Mountains. All tributaries emptying into the Mississippi have their own drainage area, forming a part of a larger basin.

Every year, the Mississippi River dumps hundreds of millions of tons of sediment into the Gulf of Mexico, widening the Mississippi Delta (Fig. 7–2) and slowly building up the land area of Louisiana and nearby states. The Gulf coastal states, from eastern Texas to the Florida panhandle, were built up with sediments eroded from the interior of the continent and hauled in by the Mississippi and other rivers. Streams, heavily laden with sediments, overflow their beds and are forced to detour as they meander toward the sea. When the streams reach the ocean, their velocity falls off sharply, and the sediment load drops out of suspension. In addition, chemical solutions carried by the rivers mix thoroughly with seawater through the action of ocean waves and currents.

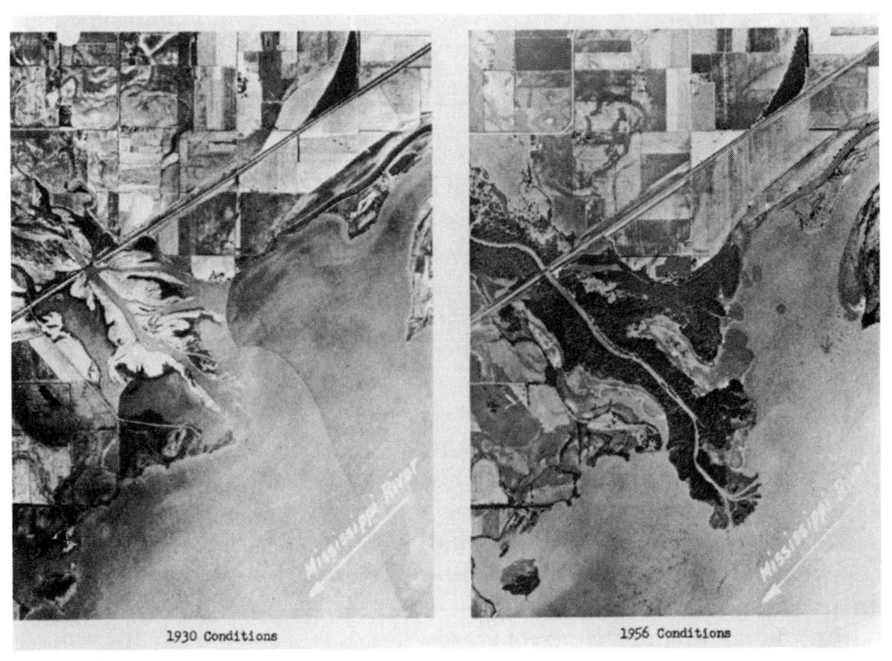

Figure 7–2 Sediment deposition in the Mississippi River delta: 1930 conditions (left), 1956 conditions (right). Photo by H. P. Guy, courtesy of USGS

TABLE 7–1 MAJOR CHANGES IN SEA LEVEL

Date	Sea Level	Historical Event
2200 B.C.	Low	
1600 B.C.	High	Coastal forest in Britain inundated by the sea
1400 B.C.	Low	
1200 B.C.	High	Egyptian ruler Ramses II builds first Suez canal
500 B.C.	Low	Many Greek and Phoenician ports built around this time are now under water
200 B.C.	Normal	
A.D. 100	High	Port constructed well inland of present-day Haifa, Israel
A.D. 200	Normal	
A.D. 400	High	
A.D. 600	Low	Port of Ravenna, Italy becomes landlocked
		Venice is built; presently being inundated by the Adriatic Sea
A.D. 800	High	
A.D. 1200	Low	Europeans exploit low-lying salt marshes
A.D. 1400	High	Extensive flooding in low-lying countries along the North Sea. The Dutch begin building dikes

Upon reaching the ocean, the riverborne sediments settle out of suspension by grain size. The coarse-grained sediments deposit near the turbulent shore and the fine-grained sediments deposit in calmer waters farther out to sea. As the shoreline advances toward the sea because of the buildup of coastal sediments or a falling sea level, finer sediments are covered by progressively courser ones. As the shoreline recedes through the lowering of the land surface or a rising sea level, coarser sediments are covered by progressively finer ones.

The difference in sedimentation rates as the sea transgresses and recedes produces a recurring sequence of sands, silts, and muds. The sands comprise quartz grains about the size of beach sands, and marine sandstones exposed in the American West were deposited along the shores of ancient seas. Gravels are rare in the ocean and move mainly from the coast to the deep abyssal plains by submarine slides. In dry regions where dust storms are prevalent, the wind airlifts fine sediment out of the region. Windblown sediments landing in the ocean slowly build deposits of red clay, whose color signifies its terrestrial origin, whereas green or gray sediments indicate a marine environment.

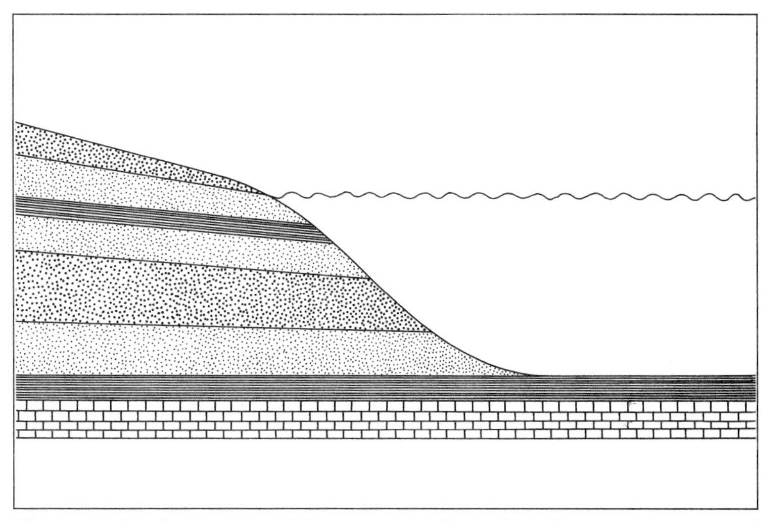

Figure 7–3 A stratigraphic cross-section showing a sequence of sandstones, siltstones, and shales overlying a basement rock composed of limestone.

The weight of the overlying sedimentary layers pressing down on the lower strata lithifies the sediments into solid rock, providing a geologic column of alternating beds of limestone, shales, siltstones, and sandstones (Fig. 7–3). Abrasion eventually grinds down all rocks to clay-size particles. Because clay particles are small and sink slowly, they normally settle out in calm, deep waters far from shore. Compaction from the weight of the overlying strata squeezes out water between sediment grains, lithifying the clay into mudstone or shale.

The varying thicknesses of sediment layers reflects the different depositional environments at the time they were laid down. Thick sandstone beds might be interspersed with thin beds of shale, indicating periods of coarse sediment deposition punctuated by periods of fine sediment deposition. Graded bedding occurs when particles in a sedimentary bed vary from coarse at the bottom to fine at the top. This type of bedding indicates the rapid deposition of sediments of differing sizes by a fast-flowing stream emptying into the sea. The largest particles settle out first and, because of the difference in settling rates, are covered by progressively finer material. Beds also grade laterally, producing a horizontal gradation of sediments from coarse to fine.

The sediments settle onto the continental shelf, which extends up to 100 or more miles and reaches a depth of roughly 600 feet. In most places, the continental shelf is nearly flat, with an average slope of only about 10 feet per mile. Beyond the continental shelf lies the continental slope, which

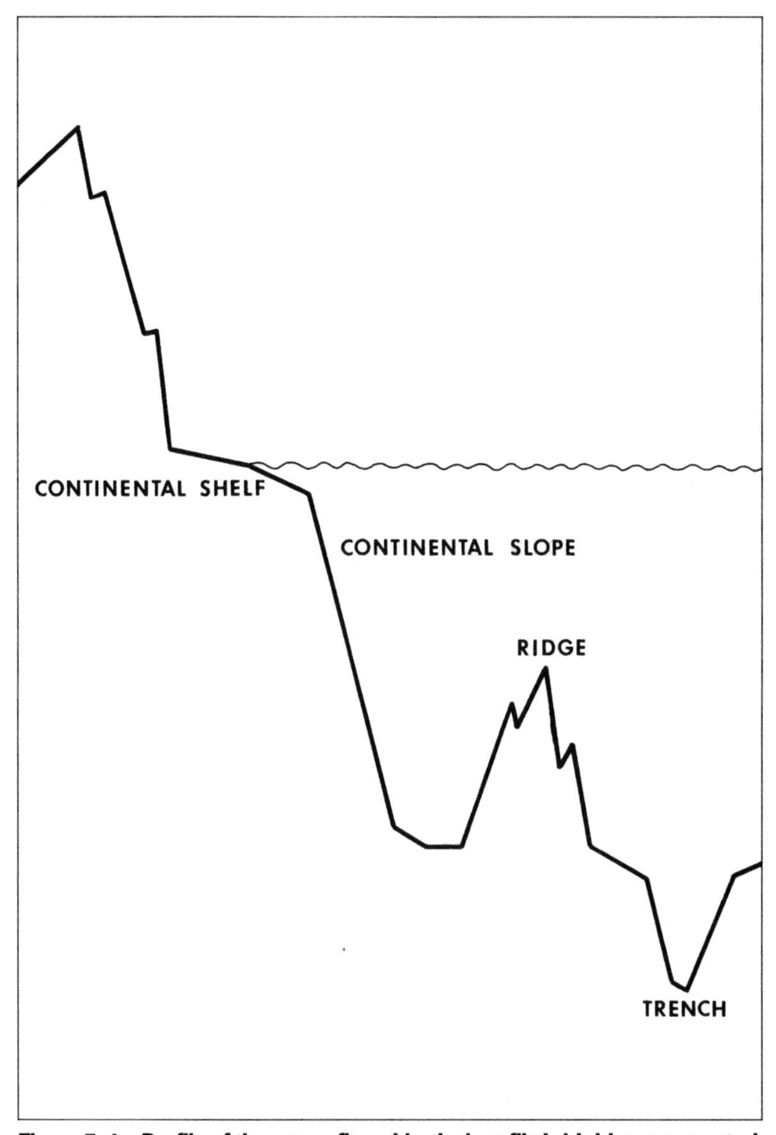

Figure 7–4 Profile of the ocean floor. Vertical profile is highly exaggerated.

extends to an average depth of more than 2 miles (Fig. 7–4). It has a steep angle, comparable to the slopes of many mountain ranges.

Sediments reaching the edge of the continental shelf slide down the continental slope under the pull of gravity. Often, huge masses of sediment cascade down the continental slope by gravity slides that can gouge out steep submarine canyons. They play an important role in building up the continental slope and the smooth ocean bottom below.

STORM SURGES

Storms at sea produce pressure changes and strong winds that pile up seawater and cause flooding when occurring at high tides. Waves generated by high winds superimposed on regular tides produce the most severe tidal floods, especially when the moon, sun, and Earth are in alignment. While

TABLE 7–2 THE BEAUFORT WIND SCALE

Beaufort Number	Description	Miles per Hour	Indications
0	Calm		Smoke rises vertically
1	Light air	1–3	Direction of wind shown by smoke drift but not by wind vane
2	Light breeze	4–7	Wind felt on face; leaves rustle
3	Gentle breeze	8–12	Leaves and small twigs in constant motion; wind extends light flag
4	Moderate breeze	13–18	Raises dust and loose vapor; moves small branches
5	Fresh breeze	19–24	Small trees begin to sway; crested wavelets form on inland water
6	Strong breeze	25–31	Large branches in motion; telephone wires whistle
7	Near gale	32–38	Whole trees in motion; resistance when walking against the wind
8	Gale	39–46	Breaks twigs off trees; large waves form on open ocean
9	Strong gale	47–54	Breaks large limbs off trees; slight structural damage occurs
10	Storm	55–63	Uproots trees; considerable structural damage occurs
11	Violent storm	64–75	Widespread damage; beach erosion occurs in coastal areas
12–17	Hurricane	>75	Devastation occurs; storm surge damages coastal areas

the tide is in, high waves raise the tide's maximum level. Strong onshore winds blowing toward the coast push seawater onto the shore. The opposite condition occurs when strong offshore winds blow toward the ocean during low tide, lowering the sea significantly and sometimes grounding vessels in port.

Most high waves and beach erosion occur during coastal storms. Thunderstorms and squalls are the most violent storms. They are most frequent in the mid-latitudes and produce gusty winds, hail, lightning, and a rapid buildup of seas. The life cycle of a single thunderstorm cell is usually less than half an hour. When the cell dies, a new one develops in its place.

Frontal storms form at the leading edge of a cold front. A squall line often precedes a cold front, with a distinctive dark gray, cylindrical-shaped cloud that appears to roll across the sky from one end of the horizon to the other (Fig. 7–5). Squall lines travel about 25 miles per hour, with winds in the squall reaching 60 miles per hour. However, they are generally short-lived, usually lasting less than 15 minutes. When a squall arrives it produces waves several feet high, but since the winds do not last long the waves die down almost as rapidly as they build up.

Figure 7–5 The leading edge of a roll cloud formation as a prefrontal squall line passes Jacksonville, Florida. Courtesy of U.S. Navy

Figure 7–6 Overwash and storm-surge penetration near Cape Hatteras, North Carolina, in 1984. Photo by R. Dolan, courtesy of USGS

Hurricanes and typhoons produce the most dramatic storm surges (Fig. 7–6). Hurricane-force winds caused by the rotation and forward motion of the storm reach 100 miles per hour or more, pushing water out in front of the storm. The low pressure in the eye of the hurricane draws water up into a mound several feet high. As the hurricane moves across the ocean and its speed matches the speed of the waves, it often sets up a resonance with the swells it generates. This adds to the height of the swells, which have been reported as more than 60 feet high in some hurricanes.

When a hurricane approaches the coast, the water piled up by the wind, the mounding of water by the low pressure, and the generation of swells and the possible resonance of swell waves can make a most deadly combination, especially when superimposed on the regular cycle of incoming tides. The result is massive flooding, devastation of property, and the loss of life.

Tidal floods are overflows on coastal areas bordering the ocean, an estuary, or a large lake. Coastal lands, including bars, spits, and deltas, offer the same protection from the sea that floodplains do from rivers. Coastal flooding is primarily a result of high tides, waves from high winds, storm surges or tsunamis, or any combination of these. Tidal floods also occur when waves generated by hurricane winds combine with flood runoff due to heavy rains that accompany the storms.

The flooding can extend over large distances along a coastline. The duration is usually short and depends on the elevation of the tide, which usually rises and falls twice daily. If the tide is in, other forces that produce high waves can raise the maximum level of the prevailing high tide. The most severe tidal floods result when waves produced by high winds combine with the regular tides, causing a tremendous amount of damage as well as severe beach erosion that continues to move the coastline inland.

COASTAL EROSION

Coastal landslides occur when a sea cliff is undercut by wave action and falls into the ocean (Fig. 7–7). Sea cliff retreat is caused by marine and nonmarine agents, including wave attack, wind-driven salt spray, and mineral solution. The nonmarine agents responsible for sea cliff erosion include chemical and mechanical processes, surface drainage water, and rainfall. Mechanical erosion processes include cycles of freezing and thawing of water in crevices, which forces existing fractures to split even farther apart. Weathering agents break down rocks or cause the outer layers to peel or spall off. Animal trails that weaken soft rock also affect sea cliff erosion, as do burrows that intersect cracks in the soil.

Surface water runoff and wind-driven rain further erode the sea cliff. Excessive rainfall along the coast also can lubricate sediments, causing huge blocks of land to slide into the ocean. Water running over the cliff edge and wind-driven rain produce the fluting often seen on cliff faces. Groundwater seeping from a cliff can create indentations on the cliff face, which undermine and weaken the overlying strata. The addition of water also increases pore pressure within sediments, reducing the shear strength (internal resistance to stress) that holds the rock together. If bedding planes, fractures, or jointing dip seaward, water moving along these areas of

Figure 7–7 Highway 1 at the Devils Slide, San Mateo County, California. Photo by R. D. Brown, courtesy of USGS

weakness might produce rock slides. This process has excavated large valleys on the windward parts of the Hawaiian Islands, where springs emerge from porous lava flows.

The main type of marine erosion is direct wave attack at the base of the sea cliff, which quarries out weak beds and undercuts the cliff until the overlying unsupported material collapses onto the beach. Waves also work along joint or fault planes to loosen blocks of rock or soil. In addition, the wind carries salt spray from breaking waves into the air and drives it against the sea cliff. Porous sedimentary rocks absorb the salty water, which evaporates, forming salt crystals that weaken rocks. The surface of the cliff slowly flakes off and falls to the beach below. The material falling to the base of the cliff piles up, forming a talas cone, a steep-sided pile of rock fragments.

Solution erosion attacks limestone cliffs, where chemical processes dissolve soluble minerals from the rocks. The seawater dissolves lime in the rocks, forming deep notches in the sea cliffs (Fig. 7–8). Chemical erosion also removes cementing agents in rocks, causing sediment grains to separate. Limestone erosion is prevalent on coral islands in the South Pacific and on the limestone coasts of the Mediterranean and Adriatic seas.

The erosion of sea cliffs and dunes that mark the coastline causes the shoreline to retreat a considerable distance. Between 1888 and 1958, the Atlantic coastline between Nauset Spit and Highland Light on Cape Cod, Massachusetts, retreated at an average rate of more than 3 feet per year. In England, the soft cliffs of the Suffolk coast, on the North Sea, erode at an average rate of 10 to 15 feet per year. At the town of Lowestoft, a single storm eroded a 40-foot-high cliff of unconsolidated rock some 40 feet. Where the cliff stood only 6 feet high, it eroded 90 feet inland.

Beach erosion is difficult to predict and almost impossible to stop. It depends on the strength of beach dunes or sea cliffs, the intensity and frequency of coastal storms, and the exposure of the coast. Most attempts to prevent beach erosion are defeated because the waves constantly batter and erode man-made defenses to keep out the sea. Jetties and seawalls erected to halt the tides increase erosion dramatically (Fig. 7–9). In their attempts to stabilize the seashores, developers are destroying the very beaches upon which they intend to build.

The rate of retreat varies with the shape of the shoreline and the prevailing wind and tides. More than half of the 72-mile-long south shore of Long

Figure 7–8 A tidal terrace at low tide, Puerto Rico. Photo by C. A. Kaye, courtesy of USGS

Figure 7–9 A system of groins to trap sand moving laterally along the beach at Lake Michigan. Photo by P. W. Koch, courtesy of USDA–Soil Conservation Service

Island, New York, is considered a high-risk zone for development, with the sea reclaiming some locations at a rate of 6 feet per year. The barrier island, running from Cape Henry, Virginia, to Cape Hatteras, North Carolina, has narrowed from both the seaward and landward sides. The rest of the North Carolina coast is moving back at 3 to 6 feet annually, and most of eastern Texas is vanishing even faster. In California, homes are falling because of the undercutting of sea cliffs (Fig. 7–10), causing considerable property damage.

About 80 percent of America's once-sandy beaches are sinking beneath the waves. Most of the problem stems from the methods that engineers use to stabilize the beaches. Jetties cut off the natural supply of sand to beaches, and seawalls increase erosion by bouncing waves back without absorbing much of their energy. The rebounding waves carry sand out to sea, undermining the beach and destroying the shorefront property the seawall was designed to protect.

In an effort to protect houses on eroding bluffs overlooking the sea, coastline residents often erect expensive seawalls. Yet these structures actually hasten the erosion of sand from the beaches in front of the wall. In effect, the seawalls are saving the bluffs to the detriment of the beaches.

Figure 7–10 The erosion of these bluffs at Point Montara, California, will eventually deliver buildings, roads, and other structures to the sea. Photo by R. D. Brown, courtesy of USGS

Barriers erected at the bottom of sea cliffs might deter wave erosion but have no effect on sea spray and other erosional processes. While beaches in front of the seawalls might lose sand naturally during certain seasons, waves return sand at other times.

Natural processes will not replenish the disappearing sand along beaches on the East Coast until the next ice age. Most of the sand currently along the coast and continental shelf originated in the north from sources such as the Hudson River. For sand to move as far south as the Carolina coast, it must progress in stages possibly taking millions of years.

As sand moves along a coast, ocean currents push it into large bays or estuaries. The embayment continues to fill with sand until sea levels drop and the accumulated sediment flushes down onto the continental shelf. In a single glacial cycle, however, the sand travels only as far as the next bay. Therefore, most beaches will not receive a major restocking of sand until the next ice age.

WAVE IMPACTS

Large storms at sea generate most ocean waves when strong winds blow across the surface of the water. Waves breaking along the coast dissipate energy and are responsible for generating along-shore currents, which in turn transport sand along the beach. Waves also cause coastal erosion, a serious problem in areas where the shoreline is steadily receding (Fig. 7–11).

Most beach erosion from high waves occurs during coastal storms. On large lakes and bays, sudden barometric pressure changes cause the water to slosh back and forth, producing a wave called a seiche. Seiches are common on Lake Michigan and on occasions can be quite destructive. Hurricanes produce the most dramatic storm surges, which are responsible for destroying entire beaches. As a wave approaches the shore, it touches bottom and slows. The shoaling of the wave distorts its shape, causing it to break upon the beach. The breaking wave dissipates its energy along the coast and causes beach erosion.

Wave reflection bounces wave energy off of steep beaches or seawalls and is responsible for the formation of sand bars. When waves approach

Figure 7–11 Beach wave erosion at Grand Isle, Louisiana. Courtesy of U.S. Army Corps of Engineers

the shore at an angle to the beach, the wave crests bend by refraction. When waves pass the end of a point of land or the tip of a breakwater, a circular wave pattern generates behind the breakwater. The refracted waves intersect other incoming waves, increasing the wave height.

Wave steepness is the ratio of wave length to wave height and is one of the most important aspects of waves. Storm waves with high steepness have short wave lengths and high wave heights and produce choppy seas. Steep waves accompanying storms at sea cause erosion of sea cliffs and sand dunes along the coast. Swells with low steepness generally result in the shoreward transport of sediment. Therefore, much of the sediment carried offshore by storm waves returns by swells during the interval between storms.

As waves leave the storm area, they develop into swells that travel great distances, sometimes halfway around the world before dying out or intercepting a coastline. As the waves spread outward from the storm area, the longer-period waves move out in front while the shorter-period waves trail behind. As swells move across the ocean toward distant shores, the low, long-period waves are the first to arrive, followed by higher swells with shorter periods.

Waves expanding outward from a storm center form rings similar to those produced by tossing a rock into a quiet pond. As the rings enlarge, the wave spreads out along a greater length, expanding the circumference of the circle. This increases the wave height as it moves away from the storm area. When swells arrive at the coast, they generate a uniform succession of waves, each with about the same period and height. The period and height change when the slower swells begin to arrive.

The wave motion changes as waves travel from deep water toward the shore. The waves transport energy but not the water itself. As the wave crest approaches, an object floating on the surface first rises and moves forward with the crest, drops into the trough, and then moves backward. Thus, a floating object describes a circular path, with the diameter equal to the wave height, and returns to its original position after the wave passes.

When swells reach a coast, they form various types of breakers (Fig. 7–12), depending on the wave steepness and bottom slope conditions near the beach. If the slope is relatively flat—less then 3 degress—the wave forms a spilling breaker, the most common type. This is an over-steepened wave that starts to break at the crest and continues breaking as the wave travels toward the beach, providing good waves for surfing.

A plunging breaker forms when the bottom slope is between 3 and 11 degrees, and the crest curls over, forming a tube of water. As the wave breaks, the tube moves toward the shore bottom and stirs up sediments. Plunging waves are the most dramatic breakers and do the most beach damage because the energy concentrates at the point where the wave breaks.

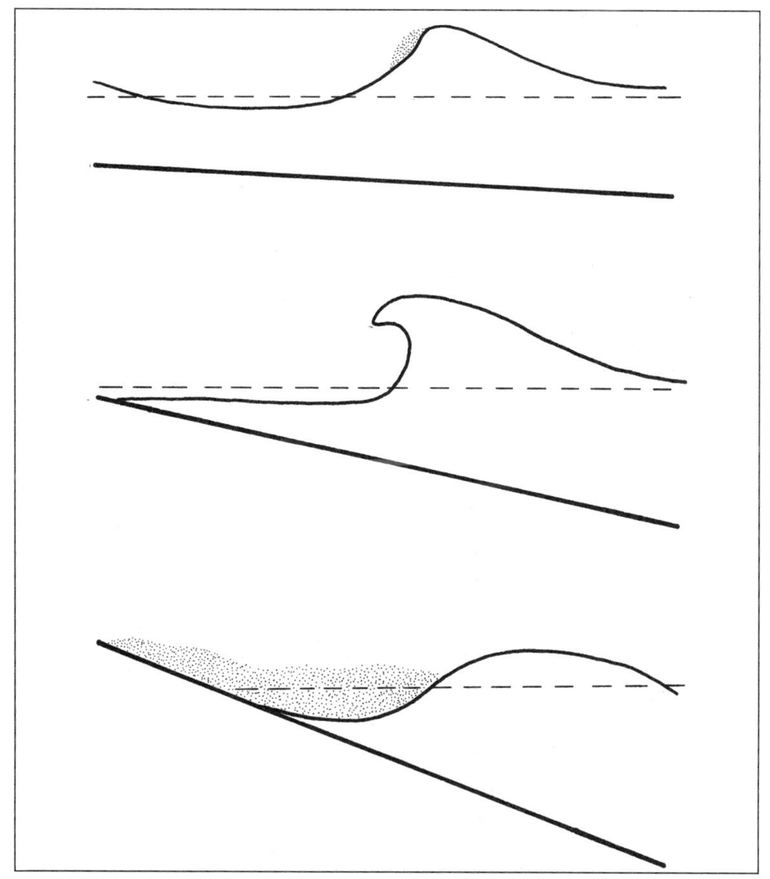

Figure 7–12 Types of breakers: spilling breaker (top), plunging breaker (middle), and surging breaker (bottom).

A collapsing breaker forms when the bottom slope is between 11 and 15 degrees. The breaker is confined to the lower half of the wave, but as the wave moves toward the coast most of it reflects off the beach.

A surging breaker develops on a steep bottom where the slope is greater than 15 degrees. The wave does not break but surges up the beach face and reflects off the coast, generating standing waves near the shore. Standing waves are important for the development of offshore structures such as bars, sand spits, beach cusps, and rip tides.

COASTAL SUBSIDENCE

Coastal subsidence often occurs during large earthquakes that cause one block of crust to drop below another. Vegetated lowlands along the coast are elevated by the influx of sediments to avoid inundation by the sea. When an earthquake strikes, these lowlands sink far enough to be submerged regularly and become barren tidal mud flats. Between earthquakes, sediments fill the tidal flats and raise them to the level where vegetation can grow once again. Therefore, repeated earthquakes produce alternating layers of lowland soil and tidal flat mud.

Earthquake-induced subsidence in the United States has occurred mainly in California, Alaska, and Hawaii. The subsidence results from vertical displacements along faults that can affect broad areas. During the March 27, 1964, Alaska earthquake, over 70,000 square miles of land tilted downward more than 3 feet, causing extensive flooding in coastal areas of southern Alaska.

Flow failures usually develop in loose saturated sands and silts on slopes with grades greater than 6 percent and originate both on land and on the seafloor near coastal areas. The Alaskan earthquake produced submarine flow failures that destroyed seaport facilities at Valdez, Whittier, and Seward. The flow failures also generated large tsunamis that overran coastal areas and caused additional damage and casualties.

Figure 7–13 Submergent coastline north of Portland, Maine. Photo by J. R. Balsley, courtesy of USGS

Some of the most dramatic examples of nonseismic subsidence in the United States are along coasts (Fig. 7–13). The Houston-Galveston area in Texas has experienced local subsidence of as much as 7.5 feet and a foot or more over an area of 2,500 miles, mostly as a result of the withdrawal of groundwater. In Galveston Bay, the ground subsided 3 feet or more over an area of several square miles following oil extraction from the underlying strata. Subsidence in some coastal towns has increased susceptibility to flooding during severe coastal storms.

The pumping of large quantities of oil at Long Beach, California caused the ground to subside, forming a huge bowl up to 25 feet deep over an area of about 20 square miles. In some parts of the oil field, land subsided at a rate of 2 feet per year. In the downtown area, the subsidence was upward of 6 feet, causing severe damage to the city's infrastructure. The injection of seawater under high pressure into the underground reservoir halted most of the subsidence, with the added benefit of increasing the production of the oil field.

Many coastal cities subside because of a combination of rising sea levels and withdrawal of groundwater, which causes compaction of the aquifer beneath the city. During the last 50 years, the cumulative subsidence of Venice, Italy, has been about 5 inches. The Adriatic sea has risen about 3.5 inches during this century, resulting in a relative sea level rise of more than 8 inches. The severe subsidence causes Venice to flood during high tides, heavy spring runoffs, and storm surges.

The overdrawing of groundwater has caused the land to sink around building foundations in the northeastern section of Tokyo, Japan. The subsidence progressed at a rate of half a foot a year over an area of about 40 square miles, a third of which sank below sea level, prompting the construction of dikes to keep out the sea from certain sections of the city. A threat of catastrophe hangs over Tokyo from inundation by floodwaters during earthquakes and typhoons that plague the region.

MARINE TRANSGRESSION

The level of the sea is rising at a rate approaching 10 times faster than only a half century ago. In most temperate and tropical regions, the sea level is rising as much as 1 inch every 5 years. Most of the increase might be due to the melting of the West Antarctic and Greeland ice sheets from two decades of apparent global warming. Ice streams flowing into the ocean calve off to form icebergs (Fig. 7–14), whose number and size seem to be increasing. In March 1995, an extremely large iceberg broke off the Antarctic ice sheet and drifted into the Pacific Ocean. In addition, alphine glaciers, which contain substantial amounts of the world's ice, appear to be melting as well.

Figure 7–14 A large iceberg along the coast of Antarctica. Courtesy of U.S. Maritime Administration

The increased temperature also causes a thermal expansion of the ocean, increasing its overall volume. Over the last century, thermal expansion has raised the level of the sea about 2 inches. Surface waters off the coast of southern California have warmed nearly 1 degree Celsius over the last half century, causing the water to expand and raise the sea level by about 1.5 inches. The additional rise in global sea levels alters the shapes of the continents and submerges low-lying barrier islands and atolls. For every foot of sea level rise, from 100 to 1,000 feet of shoreline disappears, depending on the slope of the coast.

Sea level trends are estimated from tidal gage records at stations dispersed around the world's seacoasts. In some areas, such as Louisiana, the relative level of the sea has risen as much as 3 feet per century. Louisiana is losing about 6,000 acres of land each year to the encroaching sea. The beaches along North Carolina are retreating at a rate of 4 to 5 feet per year. The higher sea levels are due in part to the sinking of the land by the increased weight of water pressing down on the continental shelf.

During the last 100 years, the global sea level has risen upward of 6 inches, due mainly to the melting of the polar ice caps. Sea ice forms a frozen band around Antarctica (Figs. 7–15a&b) and covers most of the Arctic Ocean during the winter season in each hemisphere. The total surface area of the ice appears not to have changed significantly in the last 100 years. However, the maximum extent that the ice pack reaches outward from the poles during the winter season has diminished. The ice obtains its maximum extent during the spring in the Southern Hemisphere, from

Figure 7–15a U.S. Coast Guard icebreaker *Polar Star* near Palmer Peninsula, Antarctica. Photo by E. Moreth, courtesy of U.S. Navy

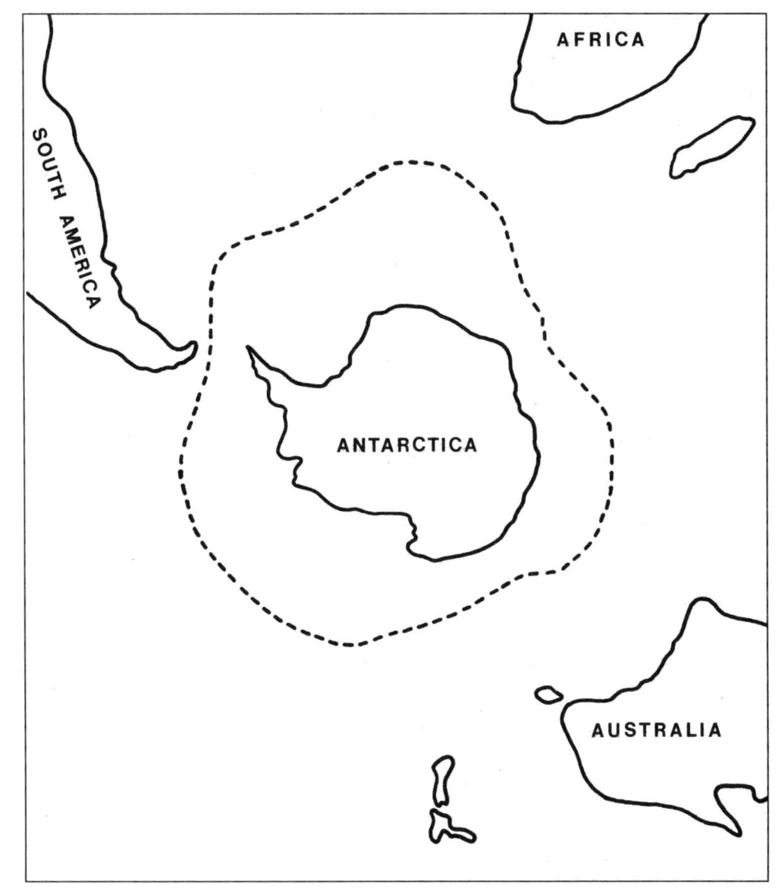

Figure 7–15b Dashed lines indicate the normal extent of sea ice around Antarctica.

October through December, when the Antarctic ice is breaking up and the Arctic ice is starting to expand. As the ocean continues to warm, the ice melts closer to the poles, further reducing the perimeter of sea ice.

During the Antarctic winter, from June through September, sea ice covers nearly 8 million square miles of ocean that surrounds the continent, with an average thickness of less than 3 feet. Because of this great expanse of ice, Antarctica plays a more significant role in atmospheric and oceanic circulation than does the Arctic. The sea ice is punctured in various places by coastal and ocean polynyas, large open-water areas that are kept from freezing by upwelling warm water currents. The coastal polynyas are like sea ice factories because they expose portions of open ocean that later freeze, continuing the ice-making process.

Most countries would feel the adverse affects of rising sea levels as rising sea temperatures cause the ice caps to melt. If the melting continues at its present rate, the sea could rise 6 feet by the middle of the next century. Large tracks of coastal land would disappear along with shallow barrier islands and coral reefs. Low-lying fertile deltas that support millions of people would drown. Delicate wetlands, where many species of marine life hatch their young, would be reclaimed by the ocean. Vulnerable coastal cities would have to relocate farther inland or build costly seawalls to protect against the rising waters.

8

SEA RICHES

The world is fortunate to have such an abundance of natural resources, which has dramatically advanced civilization. Much of this wealth comes from the sea, which holds the key to unheard of riches. Hidden in the world's oceans are untouched reserves of petroleum and minerals, along with huge fisheries that provide half the dietary protein requirements for the human race.

The capacity of the oceans to generate energy surpasses that of all available fossil fuels combined, and the harnessing of this vast energy source could meet the demand for centuries to come. New frontiers for future exploration include the continental shelves and the ocean depths. Improved exploration techniques will ensure, with proper management, a continued supply of ocean resources well into the future.

LAW OF THE SEA

The United States initiated the expansion of national claims to the ocean and its resources with the Truman Proclamations on the Continental Shelf and the Extended Fisheries Zone of 1945. Other nations followed this expansion of national boundaries and began carving up the world's oceans

TABLE 8–1 THE FUTURE OF SOME NATURAL RESOURCES

Commodity	Consumption Rate in Years	
	Reserves*	Resources
Aluminum	250	800
Coal	200	3000
Platinum	225	400
Cobalt	100	400
Molybdenum	65	250
Nickel	65	160
Copper	40	270
Petroleum	35	80

* Reserves are recoverable resources with today's mining technology

in a manner similar to the colonial division of Africa a century earlier. On December 6, 1982, 119 countries signed the United Nations Convention on the Law of the Sea. The declaration was a kind of constitution for the sea and put 40 percent of the ocean and its bottom adjacent to the coasts of continents and islands under the management of the states in possession of those regions. The other 60 percent of the ocean surface and the water below it was reserved for the traditional freedom of the seas.

The remaining wealth of the ocean floor, or about 40 percent of the Earth's surface, was deeded to the Common Heritage of Mankind. The convention placed that heritage under the management of an International Seabed Authority, with the capacity to generate income, the power of taxation, and an eminent domainlike authority over ocean-exploiting technology. The convention also provided a comprehensive global framework for protecting the marine environment, a new regime for marine scientific research, and a comprehensive legal system for settling disputes. It ensured freedom of navigation and free passage through straits used for international maritime activities, a right that cannot be suspended under any circumstances. In essence, the Law of the Sea provided a new order more responsive to the real needs of the world.

Coastal states were accorded a 12-mile limit of territorial sea and a 12-mile contiguous zone. Beyond these limits, each state was granted a 200-mile economic zone (Fig. 8–1) that includes fishing rights and rights over all resources. In cases where the continental shelf extends beyond the 200-mile limit, the economic zone with respect to resources on the seabed is expanded to 350 miles. The economic zone concept also has been described as the greatest territorial grab in history, giving coastal nations unfair advantage over landlocked ones and increasing inequality among nations.

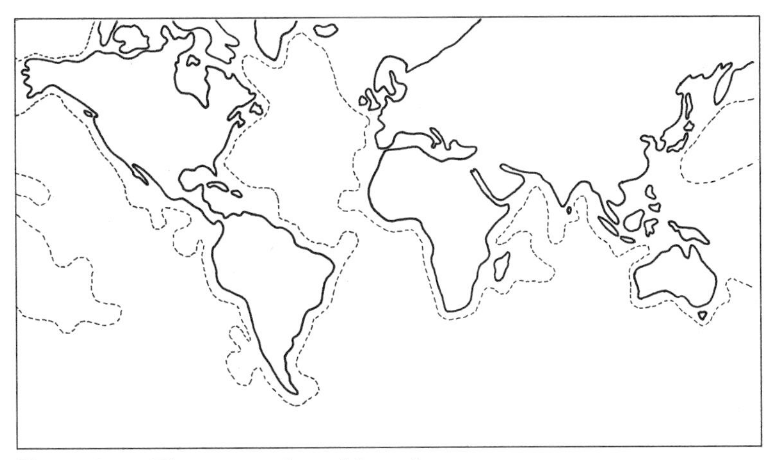

Figure 8–1 The zones of world marine resources.

In March 1983, the United States added more than 3 million square miles to its jurisdiction by declaring the waters 200 miles offshore as the nation's Exclusive Economic Zone (EEC); this area is slightly larger than the country itself. In 1984, the British oceanographic ship *Farnella* began a 6-year

Figure 8–2 The research vessel *Samuel P. Lee* carried out geophysical surveys in the Pacific Ocean and Alaskan waters. Courtesy of USGS

comprehensive mapping project of the ocean floor in the United States' EEC for future resources of petroleum and minerals. The maps revealed features possibly overlooked by smaller-scale studies. Along the West Coast were dozens of newly discovered seamounts and earthquake faults. On the western side of the Gulf of Mexico were oil-trapping salt domes, submarine slides, and undersea channels. In addition, large sand-dune fields similar to those found in the deep Pacific lay in the Gulf under 10,000 feet of water. The American research vessel *Samuel P. Lee* (Fig. 8–2) went on a similar mission in the Bering Sea to explore for oil and gas.

Figure 8–3 The seafloor drillship *Paul Langevin III* was used to obtain rock cores of the Juan de Fuca Ridge. Courtesy of USGS

While diving along a midocean spreading center called the Gorda Ridge about 125 miles off the coast of Oregon, the U.S. Navy deep submersible *Sea Cliff* discovered in September 1988 a lush community of exotic animals in a field of hot springs. Similar hot spring oases have been found on other spreading centers, where molten rock from the mantle rises to create new oceanic crust as two adjoining crustal plates pull apart. However, this was the first hydrothermal vent system existing within the United States' EEC. Moreover, the site might be a source for such strategic minerals as manganese and cobalt, used for strengthening steel. The hydrothermal water up to 400 degrees Celsius often carries dissolved minerals that form deposits on the ocean floor when the hot water mixes with near-freezing bottom water.

The discovery of a significant resource anywhere in the world's ocean could invite a claim from the nearest coastal or island state even if that resource lies beyond the limits of national jurisdiction. Such a dispute has occurred over a splattering of semisubmerged coral reefs in the South China Sea for their oil potential. Disputes over the ownership of midocean ore deposits have diminished the interests of western industrial nations, leaving the future of undersea mining and refining of manganese nodules and other metallic ores in the hands of many Asian countries, including Japan, China, South Korea, and India, which need these resources in order to reduce their dependence on foreign raw materials.

The expansion of national jurisdictions into the oceans also constrains the freedom of the seas for scientific research (Fig. 8–3). Under present law, other nations must apply for consent from a coastal state in order to conduct research in waters that were once open to all. Opposition to such a scientific project by a coastal nation that controls the waters in question might undermine the cooperative atmosphere among nations that the Law of the Sea was supposed to foster.

OIL AND GAS

Of all the mineral wealth lying beneath the waves, only oil and natural gas fields in shallow coastal waters are profitable under present economic conditions. Petroleum provides nearly half the world's energy, with about 20 percent of the oil and about 5 percent of the natural gas production offshore. In the future, perhaps half the petroleum will be extracted from under the seas. Unfortunately, much offshore oil—up to 2 million tons each year—leaks into the oceans. Such pollution could become an enormous environmental problem as production increases to keep up with demand.

Over the last 2 decades, offshore drilling for oil and natural gas in shallow coastal waters has become extremely profitable. Interest in offshore oil began in the mid-1960s, with a considerable increase in drilling a decade later following the 1973 Arab oil embargo, when American motorists stood

Figure 8–4 An oil tanker approaches the Valdez terminal of the trans-Alaskan pipeline, bringing North Slope petroleum to the lower 48 States. Courtesy of U.S. Maritime Administration

in long lines at gas stations. New important finds such as at Prudhoe Bay on Alaska's North Slope (Fig. 8–4) and on the North Sea off Great Britain resulted from intensive exploration for new reserves of offshore oil.

The desire for energy independence encouraged oil companies to explore for petroleum in the deep oceans, where they encountered many difficulties, including storms at sea and the loss of personnel and equipment. Such difficulties and problems could not justify the few discoveries that were made. Futuristic plans foresee drilling equipment and workrooms being built on the seafloor, where they are not affected by storms. This would make some deep-sea oil and gas fields available for the first time.

The creation of reservoirs of oil and natural gas requires a special set of geologic conditions, including a sedimentary source for the oil, a porous rock to serve as a reservoir, and a confining structure to trap the oil. The source material is organic carbon trapped in fine-grained, carbon-rich sediments. Porous and permeable sedimentary rock such as sandstones and limestones form the reservoir. Geologic structures produced by folding or faulting of sedimentary beds trap or pool the oil. Petroleum often associates with thick beds of salt, and because salt is lighter than the overlying sediments, it rises toward the surface, creating salt domes that help trap oil and natural gas.

The organic material that forms petroleum originates from microscopic organisms living primarily in the surface waters of the ocean and concentrated in fine particulate matter on the ocean floor. The transformation of organic material into oil requires a high rate of accumulation or a low oxygen content in the bottom water to prevent oxidation of organic material before burial under layers of sediment. Oxidation causes decay, which destroys organic matter. Therefore, areas with high rates of accumulation of sediments rich in organic material are the most favorable sites for the formation of oil-bearing rock. Deep burial in a sedimentary basin heats the organic material under high temperatures and pressures and chemically alters it. Essentially, the organic material is "cracked" into hydrocarbons by the heat generated in the Earth's interior. If the hydrocarbons are overcooked, natural gas results.

Figure 8–5 The December 19, 1976, *Argo Merchant* oil spill off Nantucket, Massachusetts. Courtesy of NOAA

The hydrocarbon volatiles locked up in the sediments along with seawater migrate upward through permeable rock layers and accumulate in traps formed by sedimentary structures that provide a barrier to further migration. In the absence of such a cap rock, the volatiles continue rising to the surface and escape into the ocean from natural seeps, amounting to about 1.5 million barrels of oil yearly. (This amount is minuscule compared to the approximately 25 million barrels of oil accidentally spilled into the ocean each year [Fig. 8–5].) Depending mainly on the temperature and pressure conditions within the sedimentary basin, it takes anywhere from several tens of millions to a few hundred million years to produce petroleum.

Reservoirs of hot gas-charged seawater called geopressured deposits lying beneath the Gulf Coast off Texas and Louisiana are a hybrid form of natural gas and geothermal energy. The gas deposits formed millions of years ago when seawater permeated porous beds of sandstone between impermeable clay layers. The seawater captured heat building up from below and dissolved methane from decaying organic matter. As more sediments piled on top of this formation, the hot gas-charged seawater became highly pressurized. Wells drilled into this formation would tap both geothermal energy and natural gas, providing an energy potential equal to about one third of all coal deposits in the United States.

The geology of the ocean floor determines whether the proper conditions exist for trapping oil and gas and greatly aids oil companies in their exploration activities. Petroleum exploration begins with a search for sedimentary structures conducive to the formation of oil traps. Seismic surveys delineate these structures by using explosions from air guns that generate waves similar to sound waves and received by hydrophones towed behind a ship (Fig. 8–6). The seismic waves reflect and refract off various

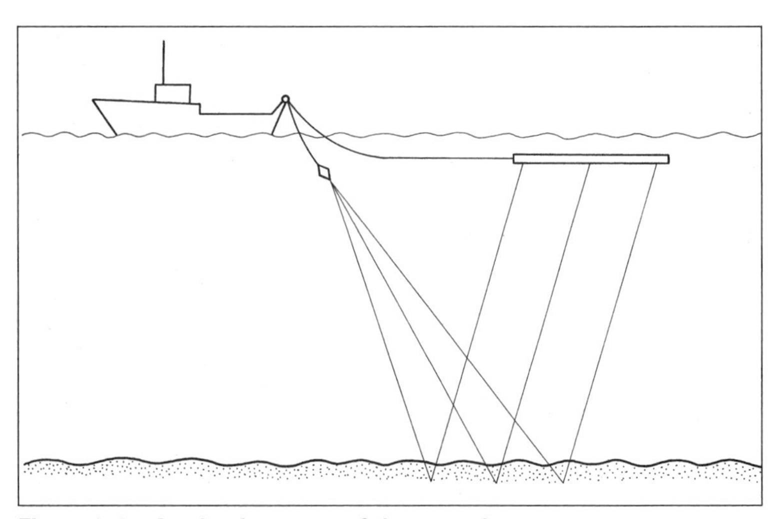

Figure 8–6 A seismic survey of the ocean's crust.

Figure 8–7 A semisubmersible drilling rig in the Mid-Atlantic outer continental shelf. Courtesy of USGS

sedimentary layers, providing a sort of geological CAT scan of the oceanic crust.

After choosing a suitable site, the oil company brings in a drill rig, which stands on the ocean floor in shallow water or free-floats anchored to the bottom in deep water (Fig. 8–7). While drilling through the bottom sediments, workers line the well with steel casing to prevent cave-ins and to act as a conduit for the oil. A blowout preventer placed on top of the casing prevents the oil from gushing out under tremendous pressure once the drill bit penetrates the cap rock. If the oil well is successful, additional wells are drilled in the area to fully develop the field.

MINERAL DEPOSITS

Hydrothermal ores deposited by hot water are associated with volcanically active zones on the ocean floor, including midocean ridges that create new oceanic crust and island arcs on the margins of subduction zones that destroy old oceanic crust. Hydrothermal deposits exist on young seafloors along active spreading centers of the major oceans as well as in regions that

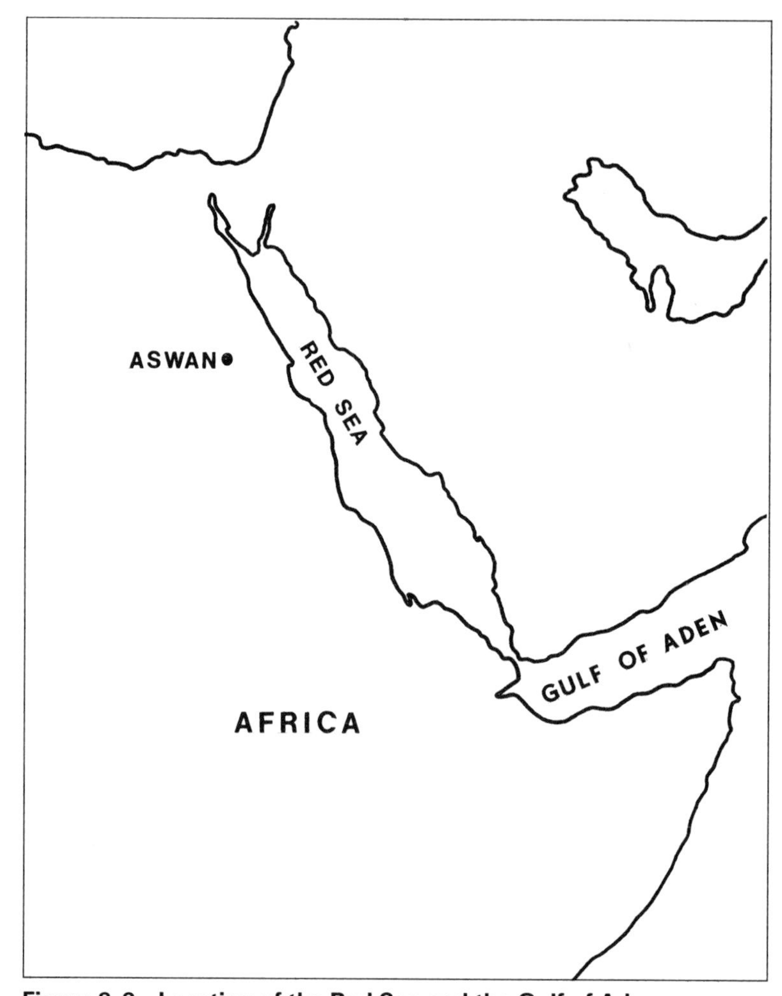

Figure 8–8 Location of the Red Sea and the Gulf of Aden.

are rifting apart and forming new bodies of water, such as the Red Sea, the Afar Rift, and the Gulf of Aden (Fig. 8–8). In addition, deep-sea drilling has uncovered identical deposits in older ocean floors far from modern spreading centers, which suggests that the process responsible for the creation of metal deposits has operated throughout the history of the major oceans.

Rich ores, including copper, zinc, gold, and silver, lie hidden among the midocean rifts. The hydrothermal deposits form by the precipitation of minerals in hot water solutions rich in silica and metals discharged from hydrothermal springs. Silica and other minerals build prodigious chimneys, from which turbulent black clouds of fluid (black smokers) billow out. Metal-rich particles precipitated from the effluent fill depressions on the seafloor and eventually form an ore body.

The minerals that contribute to hydrothermal systems originate from the mantle at depths of 20 to 30 miles below the seafloor. Magma upwelling from the mantle penetrates the oceanic crust and provides new crustal material at spreading centers. Seawater seeping into fractures in the basaltic rock on the ocean floor penetrates below the crust near the magma chamber, where it circulates within the zone of young, highly fractured rock and heats to a temperature of several hundred degrees Celsius.

The hot water kept from boiling by the pressure of several hundred atmospheres dissolves silica and minerals from the basalt, which are carried in solution to the surface by convection and discharged through fissures in the seafloor (Fig. 8–9). In addition, metal-rich fluids derived directly from the magma and volatile elements from the mantle also travel along with the hydrothermal waters to the surface. When the hot metal-rich solution emerges from a vent into cold, oxygen-rich seawater, metals such

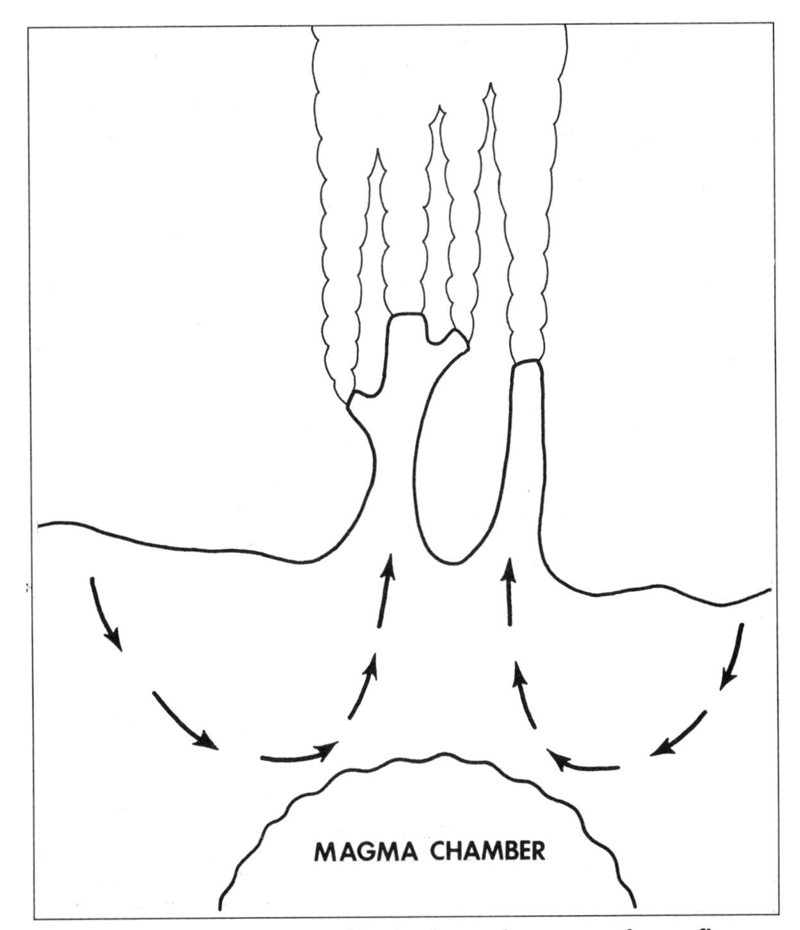

Figure 8–9 The operation of hydrothermal vents on the seafloor.

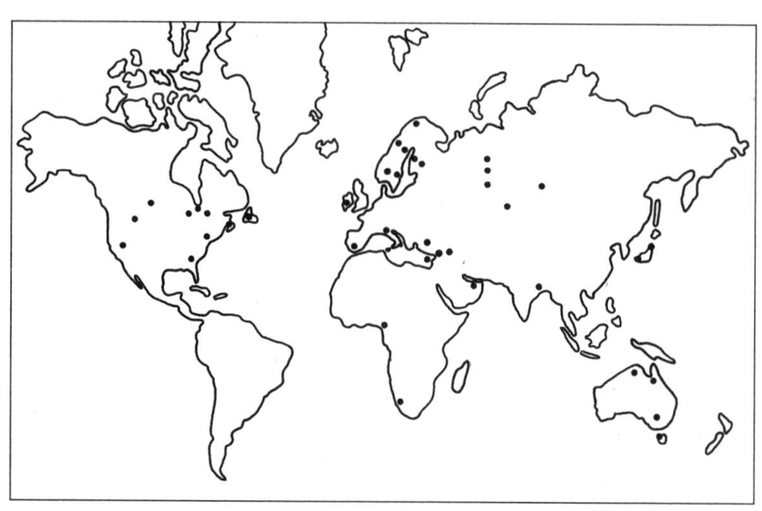

Figure 8–10 Location of ore deposits originally formed by seafloor hot springs.

as iron and manganese are oxidized and deposit along with silica. Some deposits on the Mid-Atlantic Ridge contain as much as 35 percent manganese, an important metal used in steel alloys.

The hydrothermal deposits are generally poor in copper, nickel, cobalt, lead, an zinc because these elements remain in solution longer than iron and manganese. Under oxygen-free conditions, such as those in stagnant pools of brine, copper and zinc tend to concentrate along with iron and manganese. These deposits occur in the Red Sea, where the concentrations of copper and zinc reach ore grades sufficiently high to make mining economical.

Another type of ore deposit exists in ophiolites, which are fragments of ancient oceanic crust uplifted and exposed on land by continental collisions. The grounded oceanic crust consists of an upper layer of marine sediments, a layer of pillow lava (basalts erupted undersea), and a layer of dark, dense ultramafic (iron-magnesium-rich) rocks possibly derived from the upper mantle. The metal ore deposits exist at the base of the sedimentary layer just above the area where it contacts the basalt.

Ophiolite ore deposits are scattered throughout many parts of the world (Fig. 8–10). They include the 100-million-year-old ophiolite complexes exposed on the Apennines of northern Italy, the northern margins of the Himalayas in southern Tibet, the Ural Mountains in Russia, the eastern Mediterranean (including Cyprus), the Afar Desert of northeastern Africa, the Andes of South America, on islands of the western Pacific such as the Philippines, uppermost Newfoundland, and Point Sol along the Big Sur coast of central California.

Figure 8–11a A weathered sulfide mound on the Juan de Fuca Ridge. Courtesy of USGS

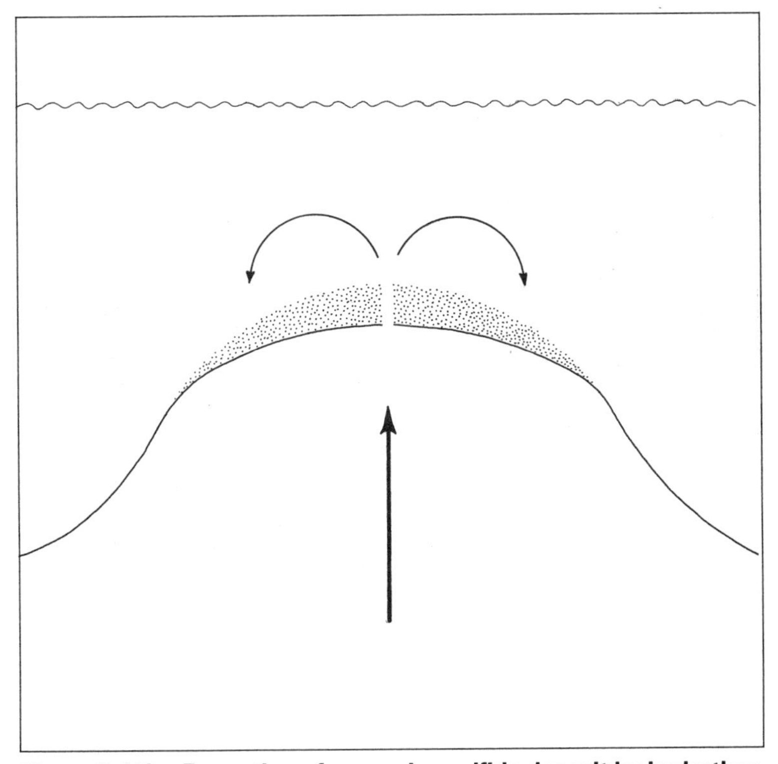

Figure 8–11b Formation of a massive sulfide deposit by hydrothermal fluids.

Massive sulfides are metal ore deposits formed at midocean spreading centers. The deposits contain sulfides of iron, copper, lead, and zinc, and occur in most ophiolite complexes, mined extensively throughout the world for their rich ores. The circulating seawater below the ocean floor acquires sulfate ions and become strongly acidic. This reaction promotes the combination of sulfur with certain metals leached from the basalt and extracted from the hydrothermal solution to form insoluble sulfide minerals.

The sulfide metals deposited by hydrothermal systems on the ocean floor form large mounds (Figs. 8–11a&b). They also occur as disseminated inclusions or veins in the rock below the seafloor in ophiolites (Fig. 8–12). Another deposit forms only when a ridge axis is near a landmass, which is a source of large amounts of erosional debris. The massive sulfide ore body lies in the midst of a sediment layer, usually shale derived from fine-grained clay. Some of the world's most important deposits of copper, lead, zinc, chromium, nickel, and platinum that are critical to modern industry originally formed several miles below the seafloor and upthrusted onto dry land during continental collisions.

Ore deposits also associate with hot brines resulting from the opening of a new ocean basin by a slow spreading center such as the one bisecting the Red Sea. Hot, metal-rich brines fill basins along the spreading zone. The cold, dense seawater percolating down through volcanic rocks becomes unusually salty because it passes through thick beds of halite (sodium chloride) buried in the crust. These salt beds are formed under dry climatic

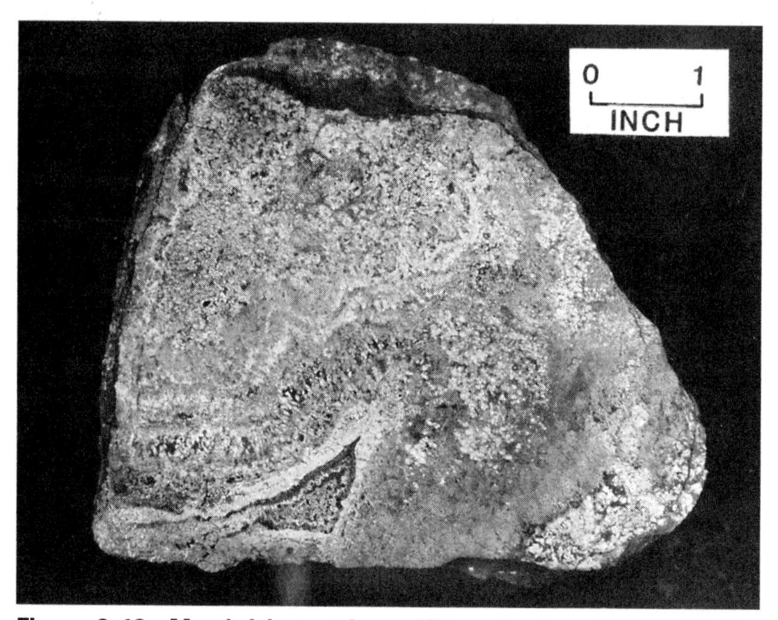

Figure 8–12 Metal-rich massive sulfide vein deposit in ophiolite.
Courtesy of USGS

Figure 8–13 Manganese nodules on Sylvania Guyot, Marshall Islands, at a depth of 4,300 feet. Photo by K. O. Emery, courtesy of USGS

conditions when evaporation exceeds the inflow of seawater in a nearly enclosed basin.

When salinity levels reached the saturation point, salt crystals precipitated out of solution and settled on the ocean floor, accumulating in thick beds. The high salinity of hot solutions circulating through these salt beds enhanced their ability to transport dissolved metals by forming complexes with the chlorine in the salt. When they discharged from the floors of the basins, the heated solutions collected as hot brines. Metals precipitate from the hot brines and settled in basins, where they formed layered deposits of metalliferous sediments up to 6 miles thick in places.

The most promising mineral deposits on the ocean floor are manganese nodules (Fig. 8–13). They are hydrogenous deposits named from the Greek words meaning "water-generated." They form on the ocean floor by the slow accumulation of metallic elements extracted directly from seawater, which contains metals such as iron and manganese in solution at concentrations of less than one part per million by weight. The metals enter the oceans from streams that transport minerals derived from the weathering and decomposition of rocks on the continents and through hydrothermal vents on the ocean floor that acquire minerals from active volcanic zones beneath the crust.

Most metallic elements have a limited solubility in an alkaline, oxygen-rich environment such as seawater. Dissolved metals such as iron and manganese are oxidized by the presence of oxygen in the seawater, forming insoluble oxides and hydroxides. The metals then deposit on the ocean floor as tiny particles or as films or crusts covering any solid material on the seafloor. Living organisms also extract certain metals from seawater; when these organisms die, their remains collect on the ocean floor, where the metals incorporate with the bottom sediments.

The growth rates of hydrogenous deposits are generally less than 1 inch in 10 million years. Most of the seafloor concretions such as manganese nodules are particularly well developed in deep, quiet waters far from continental margins and active volcanic spreading ridges where the steady rain of clay and other mineral particles prevents the metals from growing into concentrated deposits. The deposits occur in basins that receive a minimal inflow of sediments that would otherwise bury them. Depositional areas include abyssal plains and elevated areas on the ocean floor such as seamounts and isolated shallow banks.

The manganese nodules grow around a small, solid nucleus, or seed, such as a grain of sand, a piece of shell, or a shark's tooth. The seed acts as a catalyst, allowing the metals to accrete to it in a manner similar to the growth of a pearl. Concentric layers accumulate until the nodules reach about the size of a potato, giving the ocean floor a cobblestone appearance.

A ton of manganese nodules contains about 600 pounds of manganese, 29 pounds of nickel, 26 pounds of copper, and 7 pounds of cobalt. But the location of these nodules at depths approaching 4 miles makes extraction on a large scale extremely difficult. About 100 square yards of bottom ooze must be sifted to extract a single ton of nodules. One mining method would use a dredge to scoop up the nodules. Another approach would employ a gigantic vacuum cleaner to suck up the nodules. A yet more exotic scheme envisions using television-guided robots to rake up the nodules, which are crushed into a slurry and pumped to the surface.

ENERGY FROM THE SEA

The world's oceans are a large solar collector. Daily, 30 million square miles of tropical seas absorb the equivalent heat content of 250 billion barrels of oil—greater than the world's total reserves of recoverable petroleum. If only a tiny fraction of this vast store of energy is converted into electricity, it could substantially enhance the world's future energy supply. The conversion of less than a tenth of 1 percent of the heat energy stored in the surface waters of the tropics could generate roughly 15 million megawatts of electricity, or more than 20 times the current generating capacity of the United States.

Ocean thermal-energy conversion, or OTEC (Fig. 8–14), takes advantage of the temperature difference between the surface and abyssal waters. Where a significant temperature difference exists between the warm surface water and the cold deep water, efficient electrical energy can be generated. In a closed-cycle OTEC system, warm seawater evaporates a working fluid with a very low boiling point, such as freon or ammonia. The working fluid enclosed in the system recycles continuously like that in a refrigerator.

In an open-cycle OTEC system, also known as the Claude cycle after its inventor, the French biophysicist Georges Claude, the working fluid is a constantly changing supply of seawater. The warm seawater boils in a vacuum chamber, which dramatically lowers the boiling point. This system has the added benefit of producing desalinated water for irrigation in arid regions. In both systems, the resulting vapor drives a turbine to generate electricity. Cold water drawn up from depths of 2,000 to 3,000 feet condenses the gas back to a fluid to complete the cycle.

Figure 8–14 The ocean energy program at the National Renewable Energy Laboratory, Hawaii. Courtesy of U.S. Department of Energy

The nutrient-rich cold water also could be used for aquaculture, the commercial raising of fish, and serve nearby buildings with refrigeration and air conditioning. The power plant could be located onshore, offshore, or on a mobile platform out to sea. The electricity could supply a utility grid system or be used on site to synthesize substitute fuels such as methanol and hydrogen, to refine metals brought up from the seabed, or to manufacture ammonia for fertilizer.

The open-cycle system offers several advantages over the closed-cycle system. By using seawater as the working fluid, it eliminates the possibility of contaminating the marine environment with toxic chemicals. The heat exchangers of an open-cycle system are cheaper and more effective than those used in a closed-cycle system. Therefore, open-cycle plants would

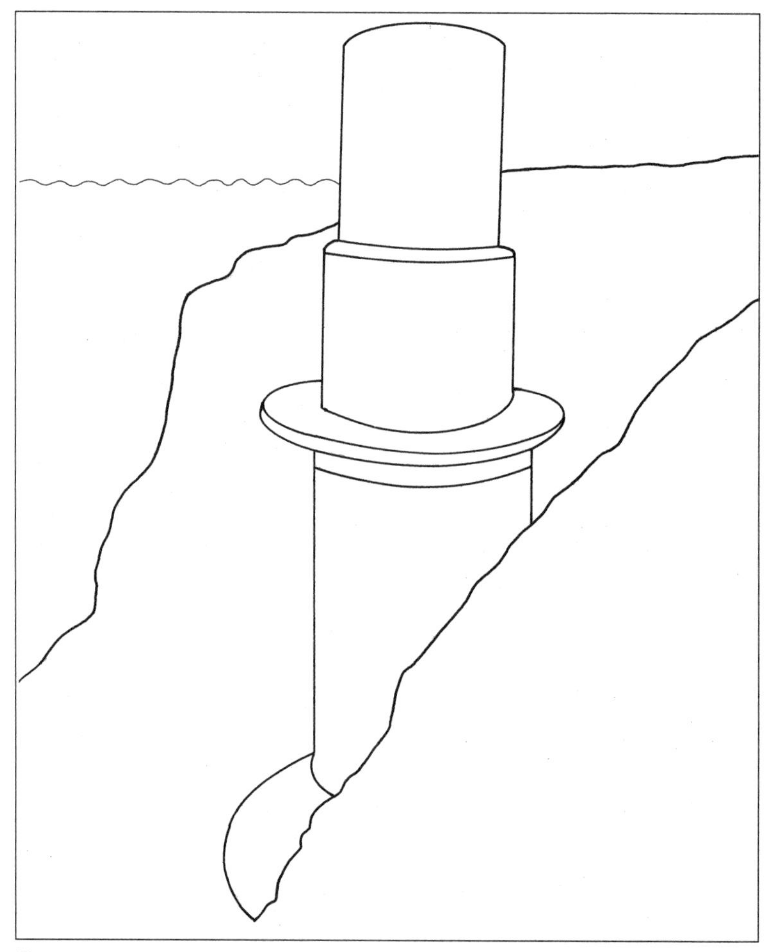

Figure 8–15 Artist's rendering of the Norsk wave-power generator on the rock coast of Bergen, Norway.

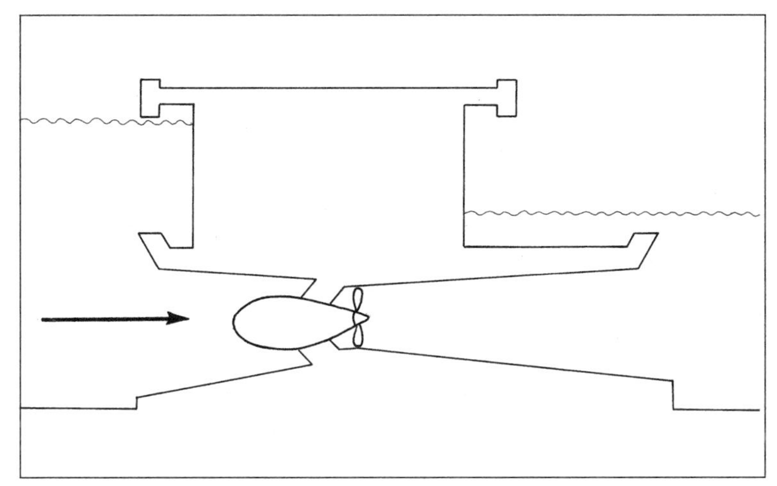

Figure 8–16 Cross-section of the La Rance tidal power station.

more efficiently convert ocean heat into electricity and would be less expensive to build.

The breaking of a large wave on the coast is a vivid example of the sizable amount of energy that ocean waves produce. The intertidal zones of rocky weather coasts receive much more energy per unit area from waves than from the sun. The waves form by strong winds from distant storms blowing across large areas of the open ocean. Local storms near the coasts provide the strongest waves, especially when superimposed on the rising and falling tides. Many hydroelectric schemes have been developed to utilize this abundant form of energy, which is economical and efficient. A crashing wave at the base of a wave-powered generator (Fig. 8–15) compresses the air at the bottom of a chamber and forces it into a vertical tower, where the compressed air spins a turbine that drives an electrical generator.

Gulfs and embayments along the coast in most parts of the world have tides exceeding 12 feet, called macrotides. Such tides are dependent on the shapes of bays and estuaries, which channel the wavelike progression of the tide and increase its amplitude. The development of exceptionally high tidal ranges in certain embayments is due to the combination of convergence and resonance effects within the tidal basin. As the tide flows into a narrowing channel, the water movement constricts and augments the tide height.

Generating electricity using tidal power involves damming an embayment, letting it fill with water at high tide, then closing the sluice gates at the tidal maximum when a sufficient head of water can drive the water turbines (Fig. 8–16). Many locations with macrotides also experience strong tidal currents, which could be used to drive turbines that rotate with both the incoming and outgoing seawater to generate electricity.

HARVESTING THE SEA

The world's fisheries are in danger of callapsing from overfishing. The once relative abundance of various species has fallen dramatically in many parts of the world. The dangers result from a constant harvest rate of a dwindling resource caused by fluctuating environmental conditions, resulting in a major decline in fish catches. The composition of the catch is also changing toward smaller fish species, and even the average size of fish within the same species is becoming smaller.

Overfishing drives populations below levels needed for competition to regulate population densities of desired species. Therefore, under heavy exploitation, species that produce offspring quickly and copiously have a relative advantage. The extent to which these changes are due to shifts in fish populations, changes in patterns of commercial fishing, or environmental effects is uncertain. What is apparent is that if present trend continues, the world's fishery could become smaller and composed of increasingly less desirable species. The world's annual fish catch is about 100 million tons, with the northwest Pacific and the northeast Atlantic yielding nearly half the total. A pronounced decline in heavily exploited fleshy fish is compensated by increased yields of so-called "trash" fish along with other small fishes. The systematic removal of large predator fish might increase annual catches of other fish species by several million tons. However, such catches would consist of smaller fish that could eventually dominate the northern latitudes, where population changes tend to be more variable and unpredictable than in the tropical regions.

Many changes in the world's fisheries are due to the strongly seasonal behavioral patterns of the fishes as well as significant differences in climate and other environmental conditions from one season to the next. Climate influences fisheries by altering ocean surface temperatures, global circula-

TABLE 8–2 PRODUCTIVITY OF THE OCEANS

Location	Primary Production tons per year of organic carbon	Percent	Total Avaliable Fish tons per year of fresh fish	Percent
Oceanic	16.3 billion	81.5	.16 million	0.07
Coastal Seas	3.6 billion	18.0	120.00 million	49.97
Upwelling Areas	0.1 billion	0.5	120.00 million	49.97
Total	20.0 billion		240.16 million	

tion patterns, upwelling currents, salinity, pH balance, turbulence, storms, and the distribution of sea ice, all of which affect the primary production of the sea. Climatic conditions could cause a shift in species distribution and loss of species diversity and quantity.

To compensate for the shortfall in marine fisheries, a variety of sea animals are raised commercially for human consumption. The shrimp, lobster, eel, and salmon raised by aquaculture account for less than 2 percent of the world's annual seafood harvest. But their total economic value is estimated at 5 to 10 times greater. The development of acquaculture and mariculture could help meet the world's growing need for food. The Chinese lead the world with more than 25 million acres of impounded water in canals, ponds, reservoirs, and natural and artificial lakes that are stocked with fish.

The food requirements of the world also might be met by cultivating seaweed and algae, which are becoming important sources of nourishment rich in vitamins. The Japanese gather about 20 edible kinds of seaweed and weekly consume about a pound of dried algae preparations per person as appetizers or deserts; they are becoming the world's leaders in the production of sea plants. The seaweed is harvested wild, and many varieties are also cultivated. When algae grows under controlled conditions, it multiplies rapidly and produces large quantities of plant material for food.

Algae crops can be harvested every few days, whereas agricultural crops grown on land require 2 to 3 months between planting and harvesting. An acre of seabed could yield 30 tons of algae a year, compared with an average of 1 ton of wheat per acre of land. The algae can be artificially flavored to taste like meat or vegetables and are highly nutritious. The ocean farm is immensely rich and can meet human nutritional needs far into the future, provided people do not turn it into a desert as they have done with so much of the land.

9

MARINE BIOLOGY

Exploration of the ocean would not be complete without a view of sea life. The riot of life in the tropical rain forests is repeated among the animals of the seafloor (Fig. 9–1). Some of the strangest creatures on Earth live on the deep ocean bottom. The most primitive species, whose ancestors go back several hundred million years, anchor to the ocean floor. The seabed hosts an eerie world comprised of tall chimneys spewing hot, mineral-rich water that supports a variety of unusual animals in the cold, dark abyss, with no counterparts found elsewhere in the sea.

BIOLOGICAL DIVERSITY

The oceans have far-reaching effects on the composition and distribution of marine life. Marine biological diversity is influenced by ocean currents, temperature, the nature of seasonal fluctuations, the distribution of nutrients, the patterns of productivity, and many other factors of fundamental importance to living organisms. The vast majority of marine species live on continental shelves or shallow-water portions of islands and subsurface rises at depths less than 600 feet (Fig. 9–2). The ecology of shallow-water

Figure 9–1 Marine life at the 100-foot depth at the Point Loma kelp beds off San Diego, California. Photo by R. Outwater, courtesy of U.S. Navy

environments also tends to fluctuate more than habitats farther offshore, which affects evolutionary development. The richest faunas live at low latitudes in the tropics, which are crowded with large numbers of highly specialized species.

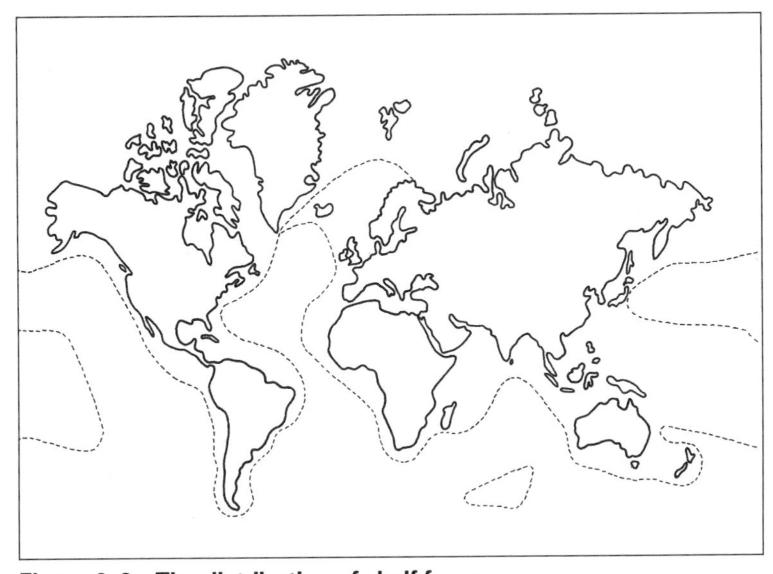

Figure 9–2 The distribution of shelf faunas.

Progressing to higher latitudes, diversity gradually falls off, until in the polar regions there are less than a tenth as many species as in the tropics. Moreover, twice as much biological diversity occurs in the Arctic Ocean, which is surrounded by continents, as in the Southern Ocean, which surrounds the continent of Antarctica. Species diversity mostly depends on the stability of the food supply. As the seasons become more pronounced in the higher latitudes, food production fluctuates much more than in the lower latitudes. Diversity is also affected by such seasonal changes as variations in surface and upwelling ocean currents that affect the nutrient supply, causing large fluctuations in productivity.

The greatest biological diversity occurs off the shores of small islands or of small continents in large oceans, where fluctuations in nutrient supplies are least affected by the seasonal effects of landmasses. The least amount of diversity is off large continents, particularly when they face small oceans, where shallow water seasonal variations are the greatest. Diversity also increases with distance from large continents. During his visit to the Galapagos Islands in the 1830s (Fig. 9–3), Charles Darwin noticed major changes in animals living on islands, compared to their relatives on adjacent continents.

Biological diversity is highly dependent on the stability of food resources, which is largely determined by the shape of the continents, the extent of inland seas, and the presence of coastal mountains. Continental platforms are particularly important, because not only do extensive shallow seas provide habitat area for shallow-water faunas, but such seas tend to

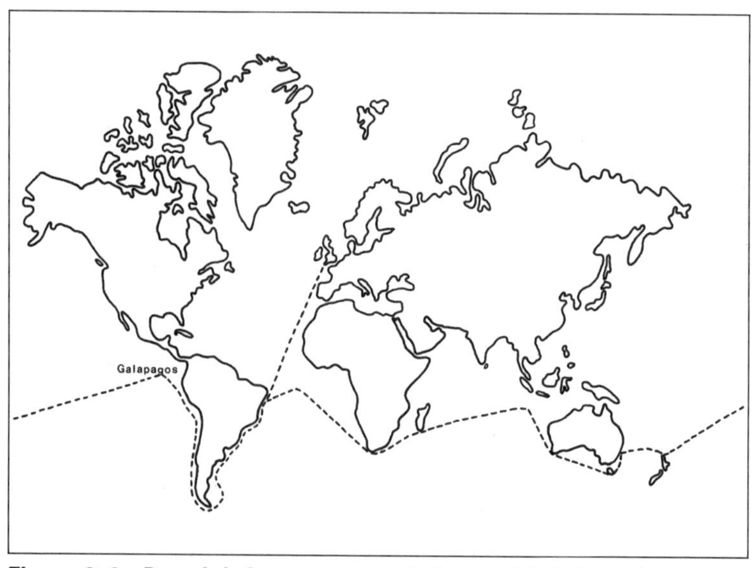

Figure 9–3 Darwin's journey around the world during his epic exploration.

dampen seasonal climatic changes and make the local environment more amenable.

Marine species living in different oceans or on opposite sides of the same ocean evolve separately from their overseas counterparts. Even along a continuous coastline, major changes in species occur that generally correspond to changes in climate because latitudinal and climatic changes create barriers to shallow-water organisms. The depth of the seafloor provides another formidable barrier to the dispersal of shallow-water organisms. Furthermore, midocean ridges form a series of barriers to the migration of marine species.

The barriers partition marine faunas into more than 30 individual "provinces." Generally, only a few common species live in each province. The shallow-water marine faunas represent more than 10 times as many species as would be present in a world with only a single province. Such single-province conditions might have occurred 200 million years ago during the existence of a single large continent and a large ocean.

The Indo-Pacific province is the widest ranging of all marine provinces and the most diverse, because of its long chains of volcanic island arcs. When long island chains align east-to-west within the same climatic zone, they are inhabited by highly diverse, wide-ranging faunas. The faunas spill over from these areas onto adjacent tropical continental shelves and islands. However, this vast tropical life is cut off from the western shores of the Americas by the East Pacific Rise, which is an effective obstruction to the migration of shallow-water organisms.

Biological diversity depends mostly on the food supply. Small, simple organisms called phytoplankton are responsible for more than 95 percent of all marine photosynthesis. They play a critical role in the marine ecology, which spans 70 percent of the Earth's surface. Phytoplankton occupy a key position in the marine food chain. They also produce 80 percent of the breathable oxygen and regulate carbon dioxide, which affects the world's climate.

The surface waters of the ocean vary markedly in color, according to the nature and amount of suspended matter such as phytoplankton, silt, and pollutants. In the open ocean, where the biomass is low, the water has a characteristic deep blue color. In the temperate coastal regions, where the biomass is high, the water has a characteristic greenish color. The temperate waters of the North Atlantic are colored green because they are rich in phytoplankton.

Upwelling currents off the coasts of continents and near the equator are important sources of bottom nutrients such as nitrates, phosphates, and oxygen. Zones of cold, nutrient-rich upwelling water scattered around the world cover only about 1 percent of the ocean but account for about 40 percent of the ocean's biological productivity and support prolific booms of phytoplankton and other marine life. These tiny organisms reside at the

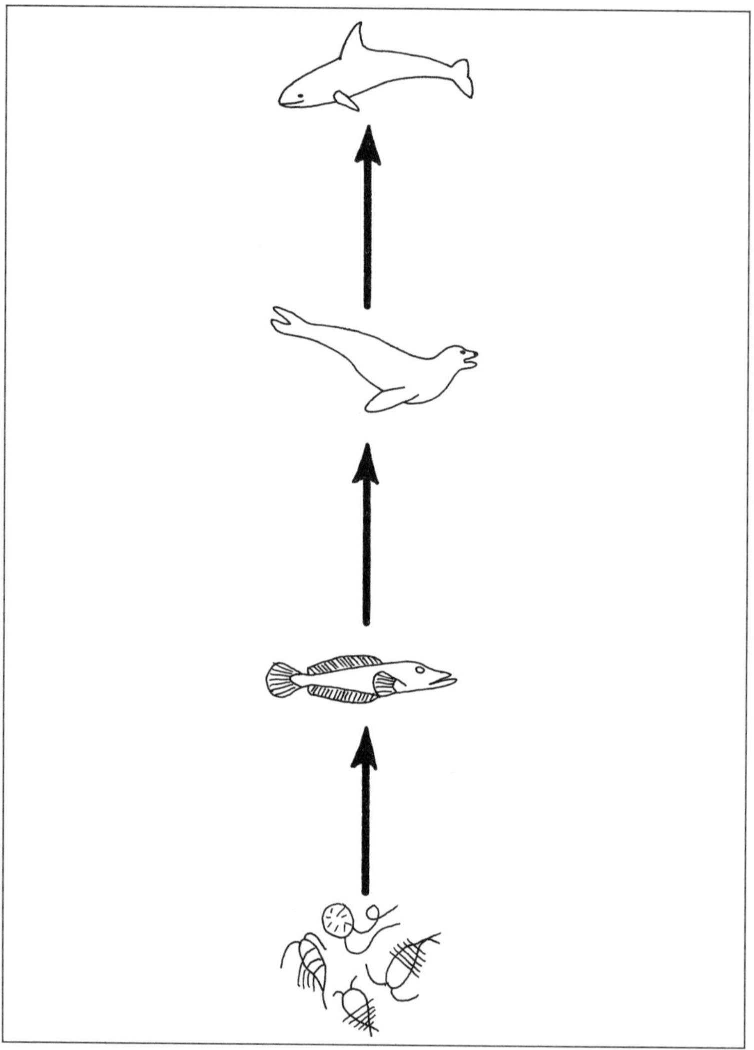

Figure 9–4 The marine food chain, from the simplest plankton to top carnivore.

very bottom of the marine food web and are eaten by predators, which are preyed upon by progressively larger predators further up the food chain (Fig. 9–4). The phytoplankton-rich ocean areas are also of vital economic importance to the commercial fishing industry.

MARINE SPECIES

The most primitive of marine species are sponges. The earliest sponges were giants as much as 10 feet or more across and grew in thickets on the

TABLE 9–1 CLASSIFICATION OF SPECIES

Group	Characteristics	Geologic Age
Protozoans	Single-celled animals. Forams and radiolarians.	Precambrian to recent
Porifera	The sponges, about 3000 living species.	Proterozoic to recent
Coelenterates	Tissues composed of three layers of cells. About 10,000 living species. Jellyfish, hydra, coral.	Cambrian to recent
Bryozoans	Moss animals. About 3000 living species.	Ordovician to recent
Brachiopods	Two asymmetrical shells. About 120 living species.	Cambrian to recent
Mollusks	Straight, curled, or two symmetrical shells. About 70,000 living species. Snails, clams, squids, ammonites.	Cambrian to recent
Annelids	Segmented body with well-developed internal organs. About 7000 living species. Worms and leaches.	Cambrian to recent
Arthropods	Largest phylum of living species, with over one million known. Insects, spiders, shrimp, lobsters, crabs, trilobites.	Cambrian to recent
Echinoderms	Bottom dwellers with radial symmetry. About 5000 living species. Starfish, sea cucumbers, sand dollars, crinoids.	Cambrian to recent
Vertebrates	Spinal column and internal skeleton. About 70,000 living species. Fish, amphibians, reptiles, birds, mammals.	Ordovician to recent

seabed. The sponge's body consists of three weak tissue layers whose cells can survive independently if separated from the main body. If a sponge is sliced up, individual pieces can grow into new sponges. Certain groups have an internal skeleton of rigid, interlocking spicules composed of calcite or silica. The great success of the sponges and other organisms such as

Figure 9–5 A helmet jellyfish under the ice of McMurdo Sound, Antarctica. Photo by W. R. Curtsinger, courtesy of U.S. Navy

diatoms, which extract silica from seawater to construct their skeletons, explains why the ocean is largely depleted of this mineral.

The coelenterates, which include corals, sea anemones, sea pens, and jellyfish (Fig. 9–5), are among the most prolific of marine animals. Most coelenterates are radially symmetrical, with body parts radiating outward from a central axis. They have a saclike body with a mouth surrounded by tentacles. The corals possess a large variety of skeletal forms, and successive generations of them have built thick limestone reefs. Corals began constructing reefs about 500 million years ago, forming chains of islands and barrier reefs along the shorelines of the continents.

The bryozoans are similar to corals and comprise microscopic individuals living in small colonies up to several inches across, giving the ocean floor a mosslike appearance. Like corals, bryozoans are retractable animals, encased in a calcareous vaselike structure, into which they retreat for safety when threatened. The polyp has a circle of ciliated tentacles, forming a net around the mouth and used for filtering microscopic food particles floating by.

The echinoderms, whose name means "spiny skin," are perhaps the strangest marine species. Their five-fold radial symmetry make them unique among the more complex animals. They are the only animals possessing a water-vascular system composed of internal canals that operate a series of tube feet or podia used for locomotion, feeding, and respiration. The great success of the echinoderms is illustrated by the fact that they have more classes of organisms than any phylum either living or extinct. The major classes of living echinoderms include starfish (Fig. 9–6), brittle stars, sea urchins, sea cucumbers, and crinoids, known as sea lilies because of their plantlike appearance.

The brachiopods, also called lamp shells due to their likeness to primitive oil lamps, were once the most abundant and diverse marine organisms. They are similar in appearance to clams, with two saucerlike shells fitted face to face that open and close using simple muscles. More advanced species called articulates have ribbed shells with interlocking teeth that maneuver along a hinge line. The brachiopods are fixed to the ocean floor with a rootlike appendage and filter food particles through opened shells that close to protect the animals against predators.

The mollusks are a highly diverse group of marine animals and finding common features among various members in often difficult. The three major groups are the snails, clams, and cephalopods. Snails and slugs comprise the largest group. The cephalopods, which include the squid, cuttlefish, octopus, and nautilus, travel by jet propulsion. They suck water into a cylindrical cavity through openings on each side of the head and expel it under pressure through a funnel-like appendage.

The nautilus (Fig. 9–7) is often referred to as a "living fossil" because it is the only living relative of the swift-swimming ammonoids. It lives in the great depths of the South Pacific and Indian oceans down to 2,000 feet. The octopus, which also lives in deep waters, is almost like an alien life form, for it is the only animal with copper-based blood, whereas all other animals have iron-based blood.

The arthropods are the largest phylum of living organisms, comprising roughly a million species, or about 80 percent of all known animals. The

Figure 9–6 A starfish off Point Loma, San Diego, California. Photo by R. Outwater, courtesy of U.S. Navy

Figure 9–7 The nautilus is the only living relative of the ammonoids.

arthropod body is segmented, with paired, jointed limbs generally present on most segments and modified for sensing, feeding, walking, and reproduction. The crustaceans are primary aquatic and include shrimp, lobsters, crabs, and barnacles.

Fish comprise over half the species of vertebrates and include the jawless fish (lampreys and hagfish), the cartilaginous fish (sharks, skates, rays, and ratfish), and the bony fish (salmon, swordfish, pickerel, and bass). The ray-finned fishes are by far the largest group of fish species. Sharks have been highly successful for the last 400 million years. They play a critical role by preying on sick and injured fish and thus help keep the ocean healthy. Closely related to the sharks are the rays, whose pectoral fins are enlarged into wings, allowing them to literally "fly" through the sea.

Life in the Abyss

The deep waters of the open ocean were once thought to be a lifeless desert. While dredging the ocean bottom in the 1870s, the British oceanographic ship *Challenger* hauled to the surface a large collection of deep-water and bottom-dwelling animals even from the deepest trenches, including hundreds of species never seen before and unknown to science. The catch comprised some of the most bizarre life forms molded by adaptative behavior and natural selection to the cold and dark of the abyss, along with several species thought to have long been extinct.

A century later, a population of large, active animals were discovered thriving in total darkness as deep as 4 miles. These depths were previously thought to be the domain of small, feeble creatures such as sponges, worms, and snails that were specially adapted to live off the debris of dead animals raining down from above. But in fact much of the deep seafloor was teeming with many species of scavengers, including highly aggressive worms, large crustaceans, deep-diving octopuses, and a variety of fishes including giant sharks.

The large physical size of many species is due to an abundance of food, lower levels of competition, and a lack of juveniles, which live in shallower water before descending to deeper depths after they mature. Large numbers of fish from the great depths of the lower latitudes are related to shallow-water varieties of the higher latitudes. Some arctic fishes might represent near-surface expressions of populations that inhabit the cold, deep waters off continental margins.

The coelacanth (Fig. 9–8), once thought to have gone extinct along with the dinosaurs and ammonoids, reemerged in 1938 when fishermen caught

Figure 9–8 Modern coelacanths have not changed significantly from their 460-million-year-old ancestors.

a 5-foot species in the deep, cold waters of the Indian Ocean off the Comoro Islands, near Madagascar. The fish looked ancient, a castaway from the distant past, with a fleshy tail, a large set of forward fins behind the gills, powerful square toothy jaws, and heavily armored scales. Stout fins enabled the fish to crawl along on the deep ocean floor in much the same manner that its ancestors crawled out of the sea to populate the land.

The oldest species living in the world's oceans thrive in cold waters. Many arctic species, including certain brachiopods, starfishes, and bivalves, belong to biological orders whose origins extend back hundreds of millions of years. Some 70 species of marine mammals known as cetaceans are among the most adaptable animals and include dolphins, porpoises, and whales, which spend much of their time feeding in the cold arctic waters of the polar regions.

The Antarctic Sea is the coldest marine habitat in the world (Fig. 9–9). It was once though to be totally barren of life. But in 1899, a British expedition to the southernmost continent found examples of previously unknown fish species related to the perches common in many parts of the world. Upward of 100 species of fish are confined to the Antarctic region, accounting for about two-thirds of the fish species in the area. Living in subfreezing waters, these fish rely on a chemical antifreeze-like substance in their bodies to prevent freeze-up during the cold Antarctic winters.

Figure 9–9 A view of marine life found on the bottom of McMurdo Sound, Antarctica. Photo by W. R. Curtsinger, courtesy of U.S. Navy

A circum-Antarctic current isolates the Antarctic Sea from the general circulation of the ocean and serves as a thermal barrier, impeding the inflow of warm currents and warm-water fishes along with the outflow of Antarctic fishes. Also, because of the extreme cold and low productivity, the Antarctic Sea is less diverse than the Arctic Ocean, which supports almost twice as many species. Biologists of the Smithsonian Institution using a deep-sea submersible made a remarkable discovery off the Bahamas in 1983. A totally new and unexpected algae lived on an uncharted seamount at a depth of about 900 feet, deeper than any previously known marine plant larger than a microbe. The species comprised a variety of purple algae with a unique structure, consisting of heavily calcified lateral walls and very thin upper and lower walls. The cells grew on top of each other, like cans stacked at a grocery store, for maximum surface exposure to the feeble sunlight. The discovery expanded our understanding of the role algae play in the productivity of the oceans, marine food chains, sedimentary processes, and reef building.

CORAL REEFS

Coral reefs are the oldest ecosystems and important land builders in the tropics, forming entire chains of islands and altering the shorelines of the continents. Over geologic time, corals have built massive reefs of limestone. The reefs are limited to clear, warm, sunlit tropical oceans such as the

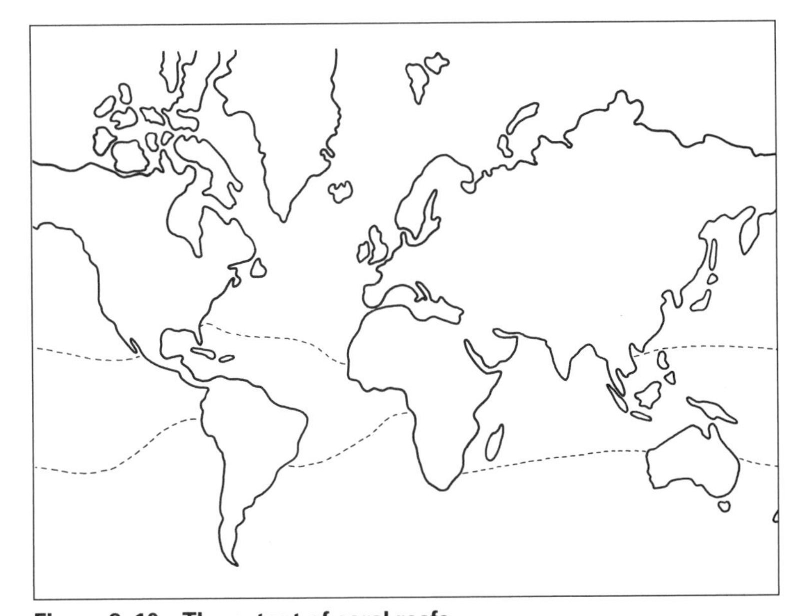

Figure 9–10 The extent of coral reefs.

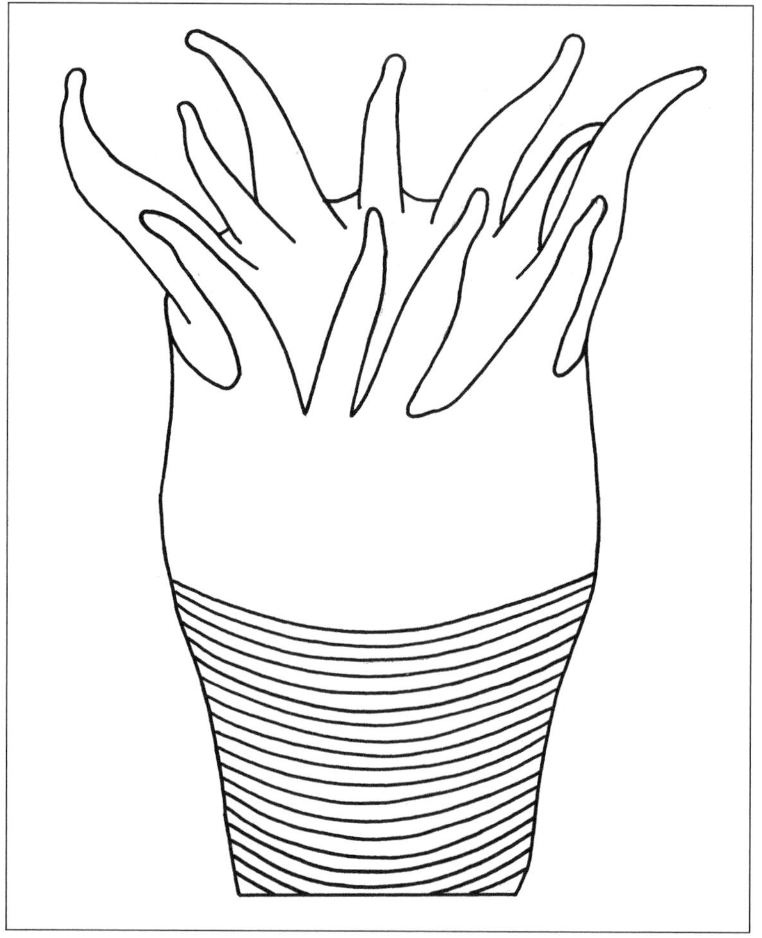

Figure 9–11 The coral polyp is a contractible animal that lives in an individual skeletal cup.

Indo-Pacific and the western Atlantic (Fig. 9–10). Hundreds of atolls comprising rings of coral islands that enclose a central lagoon dot the Pacific. The islands consist of reefs several thousand feet across, many of which formed on ancient volcanic cones that have vanished below the waves.

The coral-reef environment supports more plant and animal species than any other habitat. The key to this prodigious growth is the unique biology of corals, which plays a vital role in the structure, ecology, and nutrient cycles of the reef community. Coral reef environments have among the highest rates of photosynthesis, nitrogen fixation, and limestone deposition of any ecosystem. The most remarkable feature of coral colonies is their ability to build massive calcareous skeletons, each weighing several hundred tons.

The coral polyp (Fig. 9–11) is a soft-bodied, contractible animal, crowned with a ring of tentacles tipped with poisonous stingers that surround a mouthlike opening. The polyp lives in an individual skeletal cup, called a theca, composed of calcium carbonate. It extends its tentacles to feed at night and withdraws into the theca by day or during low tide to avoid drying out in the sun.

The corals exist in symbiosis (living together) with zooxanthellae algae within their bodies. The algae ingest the coral's waste products and produce nutrients that nourish the polyps. Since the algae need sunlight for photosynthesis, corals are restricted to warm ocean waters less than 300 feet deep,

Figure 9–12 Fire coral growing near High Cay, Andros Island, Bahamas. Photo by R. Hasha, courtesy of U.S. Navy

with much of the coral growth occurring within the intertidal zone. Widespread coral reef building indicates the presence of warm, shallow seas with little temperature variation. Dense colonies of corals mark times when the temperature, sea level, and climate were to their liking.

Coral reefs forming in shallow water where sunlight can easily penetrate for photosynthesis contain abundant organic material. Over 90 percent of a typical reef consists of fine, sandy detritus, stabilized by plants and animals anchored to the reef surface. Tropical plant and animal communities thrive on the reefs due to the coral's ability to build massive wave resistant structures. The major structural feature of a living reef is the coral rampart that reaches almost to the water's surface. It consists of massive rounded coral heads and a variety of branching corals (Fig. 9–12).

Living on this framework are smaller, more fragile corals and large communities of green and red calcareous algae. Hundreds of species of encrusting organisms such as barnacles thrive on the coral framework. Large numbers of invertebrates and fishes hide in the nooks and crannies of the reef (Fig. 9–13), some of them waiting until night before emerging to feed. Other organisms attach to virtually all available space on the underside of the coral platform or on dead coral skeletons. Filter feeders such as sponges and sea fans occupy the deeper regions.

Figure 9–13 A species of angelfish swims among the rock and coral off Andros Island, Bahamas. Photo by P. Whitmore, courtesy of U.S. Navy

Fringing reefs grow in shallow seas and hug the coastline or are separated from the shore by a narrow stretch of water. Barrier reefs also parallel the coast but are farther out to sea, are much larger, and extend for longer distances. The best example is the Great Barrier Reef, a chain of more than 2,500 coral reefs and small islands off the northeastern coast of Australia. It forms an underwater embankment more than 1,200 miles long, up to 90 miles wide, and as much as 400 feet high. It is one of the great natural wonders of the world and the largest feature built by living organisms. The Great Barrier Reef is a relatively young structure, formed largely during the Pleistocene ice ages, when sea levels fluctuated with the growth of continental glaciers during the last 3 million years.

The fore reef is seaward of the reef crest, where coral blankets nearly the entire seafloor. In deeper waters, many corals grow in flat, thin sheets to maximize their light-gathering area for photosynthesis. In other parts of the reef, the corals form massive buttresses separated by narrow sandy channels composed of calcareous debris from dead corals, calcareous algae, and other organisms living on the coral reef. The channels resemble narrow winding canyons with vertical walls of solid coral. They dissipate wave energy, allowing the free flow of sediments to prevent the coral from choking on the debris.

Below the fore reef is a coral terrace, followed by a sandy slope with isolated coral pinnacles, then another terrace, and finally a near-vertical drop off into the darkness of the abyss. The rise and fall of sea levels during the last few million years have produced terraces that resemble a stair-step growth of coral, running up the side of an island or a continent. The drowned coral represent periods of extensive glaciation, when sea levels dropped by as much as 400 feet. In Jamaica, almost 30 feet of reef have built up since the present sea level stabilized some 5,000 years ago following the last ice age.

Coral reefs are centers of high biologic productivity, sustaining fisheries that are a major source of food in the tropical regions. Unfortunately, the spread of tourist resorts along coral coasts in many parts of the world is harming the productivity of these areas due mostly to an increase in sedimentation. These developments are usually accompanied by increased sewage dumping, overfishing, and physical damage to the reef from construction, dredging, dumping, and landfills, along with the direct destruction of the reef for souvenirs and curios.

In areas like Bermuda, the Virgin Islands, and Hawaii, development and sewage outflows have caused extensive overgrowth and death of the reef by thick mats of algae. The algae suffocate the coral by supporting the growth of oxygen-consuming bacteria, particularly in the winter, when the algal cover on shallow reefs is high. This action results in the death of the living coral and the eventual destruction of the reef and the biological communities it supports.

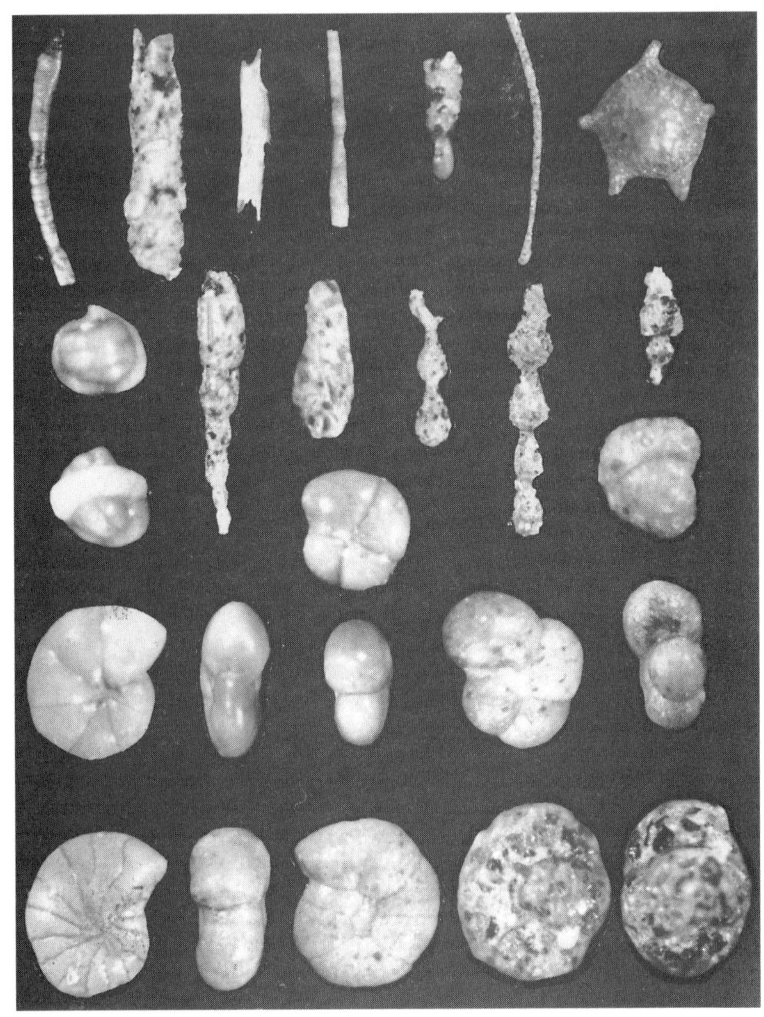

Figure 9–14 Fossil foraminirfera of the North Pacific Ocean. Photo by B. P. Smith, courtesy of USGS

Increasing ocean temperatures induced by a possible global warming are bleaching many reefs, causing the corals to turn ghostly white due to the expulsion of algae from their tissues. The algae aid in nourishing the corals, and their loss poses a great danger to the reefs. Foraminifera (Fig. 9–14), tiny marine organisms whose skeletons preserve much of the record of behavior of the ocean and climate, exhibit damage similar to that observed in bleached coral reefs. Along with corals, they play an important role in the global ecosystem wherein bleaching could seriously affect the marine food chain.

Figure 9–15 Hydrothermal vents on the deep ocean floor provide nourishment and heat for bottom-dwellers. Courtesy of USGS

THE VENT CREATURES

On the crest of the East Pacific Rise south of Baja, California, 8,000 feet beneath the ocean lies an eerie world that time forgot. In volcanically active fields, a habitat unlike any other on Earth contains species previously unknown to science that thrive in total darkness near hydrothermal vents. The undersea geysers build forests of exotic chimneys, called black smokers, that spew out hot water blackened with sulfur compounds (Fig. 9–15).

The black smokers support some of the world's most bizarre biology. Flourishing among the hydrothermal vents are perhaps the strangest animals ever encountered. Life might even have originated around such vents, obtaining from the Earth's hot interior all the necessary nutrients to survive. In such an environment, the evolution of life could have begun as early as 4.2 billion years ago.

Seawater seeping downward near magma chambers acquires heat and minerals and is expelled through fissures in the ocean floor. The hydrothermal vents not only maintain the bottom waters at livable temperatures, upward of 20 degrees Celsius, they also provide valuable nutrients, making this the only environment totally independent of the sun as a source of energy, which instead comes from the Earth itself.

Clustered around the hydrothermal vents are large communities of unusual animals (Fig. 9–16) as crowded as the tropical rain forests. Large white clams up to a foot long and mussels, having no need for skin pigments, nestle between black pillow lava. Giant white crabs scamper blindly across the volcanic terrain, and long-legged marine spiders and tiny octopuses roam the ocean floor. Living in total darkness, these species have no need for eyes, which have become useless appendages. Clusters of giant tubeworms up to 10 feet tall sway in the hydrothermal currents. The tubeworms are contractable animals that live inside long, white stalks up to 4 inches wide. While feeding, the tubeworm exposes a long bright red plume abundantly supplied with blood that is also a delicacy for the crabs, which climb the stalks to obtain a meal.

In the Atlantic, the vents are dominated by swarms of small shrimps. They originally were thought to be blind, until it was later discovered that the shrimps had an unusual pair of eyes on their backs instead of in the front of their heads. Apparently they can see using the feeble light emanating from the hot water chimneys. Biologists are intrigued by the deep-sea light because of the possibility that organisms can harness this energy using a type of photosynthesis totally independent of the sun.

Most remarkable is the fact that the vent animals do not obtain their nutrition in the form of detrital material falling from above, but rely on symbiosis with sulfur-metabolizing bacteria that live within the host's tissues. The bacteria metabolize sulfur compounds in the hydrothermal water by chemosynthesis. They harness energy liberated by the oxidation of hydrogen sulfide from the vents to incorporate carbon dioxide for the production of organic compounds such as carbohydrates, proteins, and lipids. The byproducts of the bacteria's metabolism are absorbed into the host animal and nourish it.

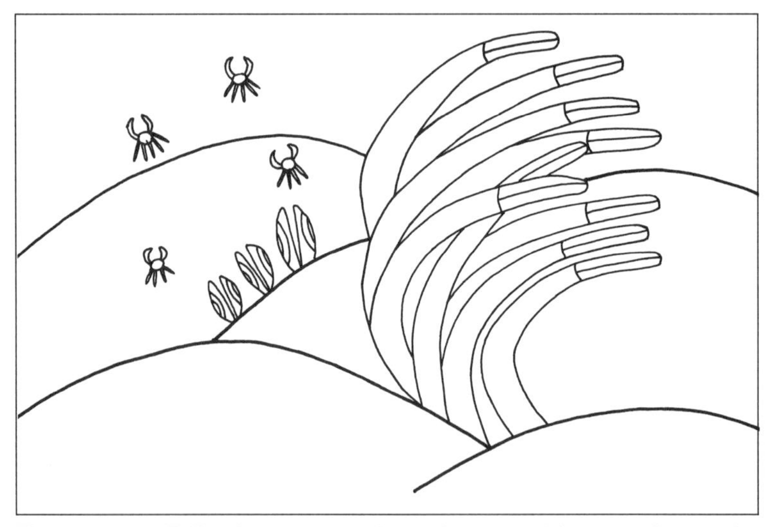

Figure 9–16 Tall tube worms, giant clams, and large crabs on the deep-sea floor near hydrothermal vents.

Some animals also feed on bacteria directly. Odd-looking colonies of bacteria—some with long tendrils that sway in the warm currents—become the feeding grounds for complex forms of higher life. Some regions are clouded by drifting bits of whitish bacteria that swirl like falling snow. Occasionally, clumps of waving bacteria break loose from fissures and join the swirl of biologic "snowfall." The vent animals are so dependent on the bacteria that the mussels have only a rudimentary stomach and the tube-worms lack even a mouth.

The animals live precarious lives. The hydrothermal vent systems turn on and off sporadically, and species can only survive as long as the vents continue to operate—perhaps for only a few years. To illustrate this point, isolated piles of empty clamshells bear witness to local mass fatalities. In new basalt fields, the vent creatures soon establish residency around young hydrothermal vents, and the once-barren abyss its rapidly colonized.

THE INTERTIDAL ZONE

The constant waxing and waning of the tides is responsible for the prodigious growth in the intertidal zone, the habitat area between high and low

Figure 9–17a Intertidal exposure of chaotic blocks of sandstone near Piller Point, Clallan County, Washington. Photo by W. O. Addicott, courtesy of USGS

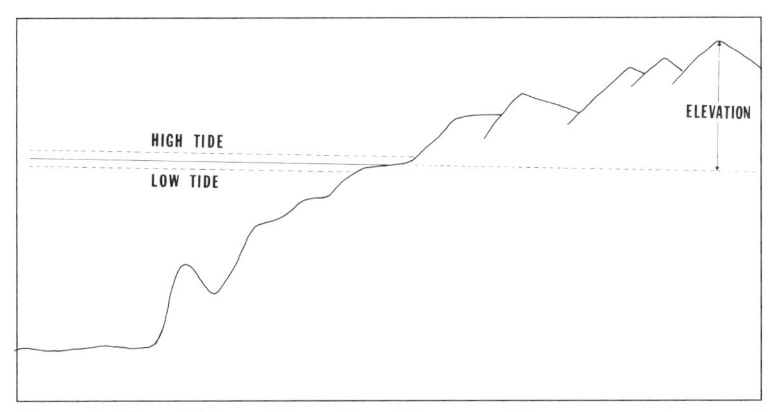

Figure 9–17b High and low tide with respect to mean sea level and elevation.

tides (Figs. 9–17a&b). Most inhabitants of the intertidal zone have biological clocks set to the rhythm of the lunar day. The rhythms are characterized by repetitive behavioral or physiological events, such as feeding, that are synchronized with the tides. Each lunar day, which is about 25 hours long, generally has two tides, producing bimodal lunar-day rhythms, as compared to the unimodal solar-day rhythms of organisms attuned to the 24-hour solar day.

Biological clocks are important aids to survival by giving advance warning of regular changes in the environment, such as nightfall or the return of the tides. Even under constant laboratory conditions without the effects of diurnal or tidal cycles, biological clocks continue to function and the tidal rhythms persist for some time.

Apparently, the tidal rhythms are not learned or impressed on the organisms by the tides themselves. Crabs raised in the laboratory and exposed only to diurnal conditions exhibit a distinct tidal component in their activity after their body temperatures are lowered. Also, crabs taken from areas not subject to tides and moved to a tidal flat quickly establish a tidal rhythm. Apparently the clock that measures the tidal frequency is innate, and only needs to be activated by an outside stimulus.

Rhythmic behavior in organisms is also an expression of the genetic code. Heredity decides whether an animal is active during high or low tide. The environment also plays an important role in the establishment of a tidal rhythm. The schedule of the tides only determines the setting of the biological clock. Therefore, animals transported to a different ocean synchronize their clocks to the new tidal conditions. Moreover, the pounding surf shapes the activity patterns of inhabitants living on beaches exposed to the open sea.

Intertidal organisms living in protected bays are not as exposed to the vicissitudes of the sea and are controlled by more subtle conditions, such as a drop in temperature or pressure changes induced by the incoming tides, which help set their tidal rhythms. Even in the absence of outside stimuli, the biological clock continues to run accurately but no longer controls the organism's activity. It operates independently from tidal influences until the organism returns to the sea and the clock takes over again. Like all clocks, the accuracy of biological clocks is not altered by changes in the environment, nor do they pertain to intertidal organisms alone but also to the entire spectrum of life.

10

RARE SEAFLOOR FORMATIONS

The floor of the ocean is host to a myriad of unique geologic structures found nowhere else on Earth. Unusual seamounts associated with deep-sea trenches erupt cold mud instead of hot lava. Scattered around the midocean ridges are remarkable volcanic deposits, including piles of pillow lavas, forests of black smokers, and undersea geysers that eject vast quantities of hot water rising toward the surface in massive plumes.

Submarine slides larger than any landslide gouge out deep chasms in the ocean floor and deposit enormous heaps of sediment on the bottom of the sea. The active seafloor sports a variety of depressions, including sea caves, blue holes, gas blowouts, calderas, and numerous craters formed by undersea explosions. Large meteorites or comets falling into the sea blast deep craters in the ocean floor, many of which are better preserved than their landward counterparts.

MUD VOLCANOES

In the Pacific Ocean, about 50 miles west of the Mariana Trench, the world's deepest depression, lies a cluster of large seamounts 2.5 miles below the surface of the sea in a zone about 600 miles long and 60 miles wide. The undersea mountains were built not by hot volcanic rock like most Pacific seamounts but by cold, plastic serpentine, which is a soft, mottled green rock similar to the color of a serpent, hence its name. Serpentine is a low-grade metamorphic rock and the main mineral in asbestos. It originates from the reaction of water with olivine, an olive-green, iron- and magnesium-rich silicate and a major constituent of the upper mantle.

The erupting serpentine rock flows down the flanks of the seamounts like lava from a volcano and forms gently sloping structures. Many of these seamounts rise more than a mile above the ocean floor and measure as much as 20 miles across at the base, resembling broad shield volcanoes such as Mauna Loa, which built the island of Hawaii. Drill cores taken during the international Ocean Drilling Program in 1989 showed that serpentine not only covered the tops of the seamounts but also filled the interiors.

Several smaller seamounts only a few hundred feet high are mud volcanoes, looking something like those in hydrothermal areas on land (Fig. 10–1). They are composed of mounds of remobilized sediments formed in association with hydrocarbon seeps, where petroleumlike substances ooze out of the ocean floor. Apparently, sediments rich in planktonic carbon are "cracked" into hydrocarbons by the heat of the Earth's interior. Even cores recovered around hydrothermal fields smell strongly of diesel fuel.

The mud comprises peridotite that is converted into serpentine and ground down into rock flour by movement along underlying faults called fault gouge. The mud volcanoes appear to undergo pulses of activity,

Figure 10–1 Mud volcanoes and acidulated ponds northwest of Imperial Junction, Imperial County, California. Photo by Mendenhall, courtesy of USGS

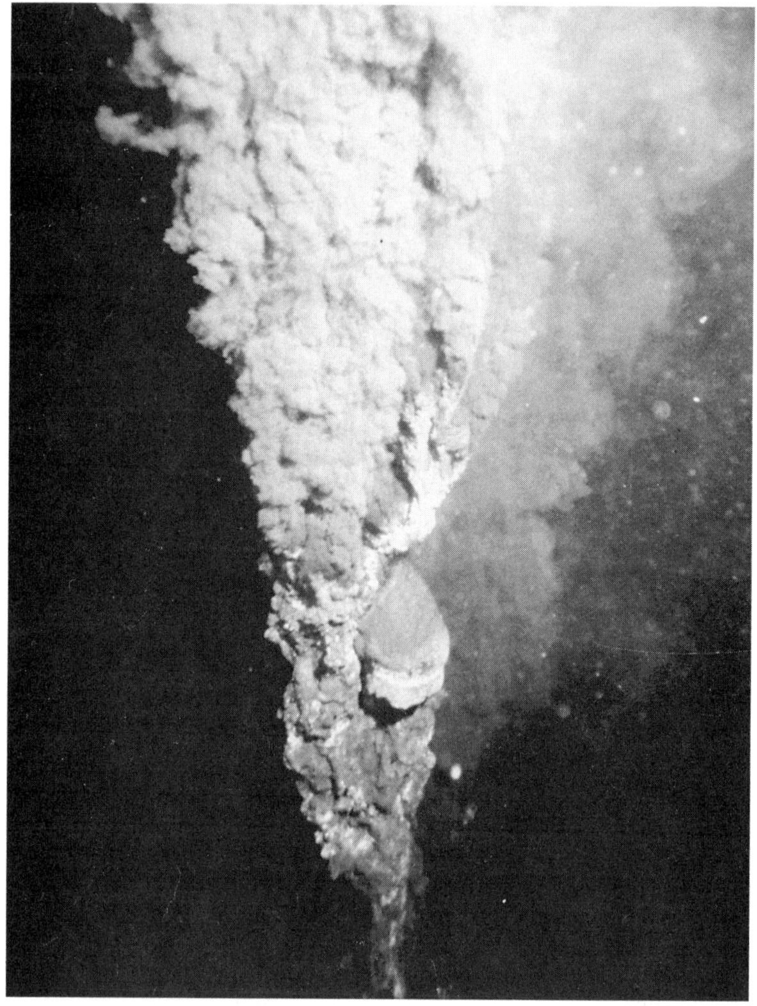

Figure 10–2 A black smoker on the East Pacific Rise. Photo by R. D. Ballard, courtesy of WHOI

interspersed with long dormant periods. Many seamounts formed recently (in geologic parlance), probably within the last million years or so.

The Mariana seamounts might be diapirs similar to the salt diapirs of the Gulf of Mexico, which trap oil and gas. The diapirs appear to be composed of the mantle rock peridotite altered by interaction with fluids distilled from the subducted portion of the Pacific plate as it descends into the Mariana Trench and slides under the Philippine plate. Fluids expelled from the subducting plate react with the mantle rock, transforming portions of the mantle into low-density minerals that rise slowly through the subduction zone to the seafloor.

About 90 million years ago, the Mariana region forward of the island arc consisted of midocean ridge and island arc basalts that have been eroding away by as much as 40 miles by plate subduction over the last 50 million years. The seamount-forming process has been proceeding for perhaps 45 million years, as oceanic lithosphere vanishes into the subduction zone, distilling enormous quantities of fluids from the descending plate. The fluids reacting with the surrounding mantle produce blobs of serpentine that rise to the surface through fractures in the ocean floor.

The fluid temperatures in subduction zones are cool, compared to those associated with midocean ridges where hydrothermal vents eject high-temperature black effluent (Fig. 10–2). But instead of comprising heavy minerals like the black smokers of the East Pacific Rise and other midocean ridges, the ghostly white chimneys of the Mariana seamounts are composed of aragonite, a calcium carbonate mineral that normally dissolves in seawater at these great depths. Hundreds of corroded and dead carbonate chimneys are strewn across the ocean floor in wide "graveyards."

Apparently, cool water emanating from beneath the seafloor surface allows the carbonate chimneys to grow and avoid dissolution by seawater. Many carbonate chimneys are thin, and they are generally less than 6 feet high. Other chimney structures are thicker and taller and occasionally coalesce to form ramparts encrusted with a black manganese deposit. Small manganese nodules are also scattered atop many of the mountains of mud.

Exotic terranes are fragments of oceanic lithosphere originating from distant sources and exposed on the continents and islands in zones where plates collide. Many terranes contain large serpentine bodies that are similar in structure to the Mariana seamounts. Their presence is a constant reminder that the ocean floor was highly dynamic in the past and continues to be so today.

SUB-SEA GEYSERS

Perhaps the strangest environment on Earth lies on the ocean floor in deep water near seafloor spreading centers such as the crests of the East Pacific Rise and the Mid-Atlantic Ridge, which are portions of the Earth's largest volcanic system. Solidified lava lakes hundreds of feet long and up to 20 or more feet deep probably formed by rapid outpourings of lava. In places, the surface of a lava lake has caved in, forming a collapse pit (Fig. 10–3).

Seafloor spreading is often described as a wound that never heals as magma slowly bleeds from the mantle in response to diverging lithospheric plates. During seafloor spreading, magma rising out of the mantle solidifies on the ocean floor, producing new oceanic crust. At the base of jagged basalt cliffs is evidence of active lava flows and fields strewn with pillow lava formed when molten rock ejects from fractures in the crust and quickly cools by the deep cold water.

Figure 10–3 The rim of a lava lake collapse pit on the Juan de Fuca Ridge in the east Pacific. Courtesy of USGS

Lava erupting from undersea volcanoes continuously forms new crust along the midocean ridges, as lithospheric plates on the sides of the rift inch apart and molten basalt from the mantle slowly rises to fill the gap. Occasionally, a colossal eruption of lava along the ridge crest flows downslope for more than 10 miles. Most of the time, however, the basalt

Figure 10–4 Cluster of tube worms and sulfide deposits around hydrothermal vents near the Juan de Fuca Ridge. Courtesy of USGS

just oozes out of the spreading ridges, forming a variety of lava structures on the ocean floor.

The ridge system exhibits many uncommon features, including massive peaks, sawtooth ridges, earthquake fractured cliffs, deep valleys, and lava formations of every description (Fig. 10–4). Lava formations associated with midocean ridges consist of sheet flows and pillow, or tube flows. Sheet flows are more common in the active volcanic zone of fast spreading ridge

Figure 10–5 Pillow lava on the ocean floor. Courtesy of WHOI

segments like those of the East Pacific Rise, where the plates are separating at a rate of 4 to 6 inches a year.

Pillow lavas (Fig. 10–5) erupt as though basalt were squeezed out of a giant tooth paste tube. They generally arise from slow spreading centers, such as those of the Mid-Atlantic Ridge, where plates spread apart at a rate of only about an inch per year, and the lava is much more viscous. The surface of the pillows often contains corrugations or small ridges, indicating the direction of flow. The pillow lavas typically form small, elongated hillocks, pointing downslope.

In rapidly spreading rift systems such as the East Pacific Rise south of Baja, California, hydrothermal vents built prodigious forests of exotic chimneys up to 30 feet tall.

Figure 10–6 Location of the East Pacific Rise.

Figure 10–7 An artist's rendition of the deep-submersible *Alvin*. Courtesy of U.S. Navy

Seawater seeping through the oceanic crust acquires heat near magma chambers below the rifts and is expelled with considerable force through vents as though they were undersea geysers. (The term *geyser* originates from the Icelandic word *geysir*, meaning "gusher.")

The hydrothermal water is up to 400 degrees Celsius but does not boil because, at this great depth, it is under a pressure of 200 to 400 atmospheres. The superhot water is rich in dissolved minerals, such as iron, copper, and zinc, that precipitate out upon contact with the cold water of the abyss. The sulfide minerals ejected from hydrothermal vents build tall chimney structures, some with branching pipes. The black sulfide minerals drift along in the ocean currents like thick effluent from factory smokestacks.

About 750 miles southwest of the Galapagos Islands, along the undersea mountain chain that comprises the East Pacific Rise (Fig. 10–6), lies an immense lava field that recently erupted. The eruption appears to have started near the ridge crest and flowed downslope over cliffs and valleys for more than 12 miles. The volume of erupted material was nearly 4 cubic miles spread over an area of some 50,000 acres, which is about half the total annual production of new basalt on the seafloor worldwide. It is enough lava to pave the entire U.S. interstate highway system to a depth of 30 feet. Although not the greatest eruption in geologic history, this could well be the largest basalt flow in historic times. Associated with these huge bursts of basalt are megaplumes of warm, mineral-laden water, measuring up to 10 miles or more across and thousands of feet deep.

The submersible *Alvin* (Fig. 10–7), launched from the oceanographic research vessel *Atlantis II*, is the workhorse for exploring the deep ocean floor. In April 1991, oceanographers aboard *Alvin* witnessed an actual eruption or its immediate aftermath on the East Pacific Rise about 600 miles southwest of Acapulco, Mexico. The scientists realized that the seafloor had recently erupted because the scenery did not match photographs taken at the same location 15 months earlier.

The scene showed recent lava eruptions that sizzled a community of tubeworms and other animals living on the deep ocean floor 1.5 miles below the sea. Suspended particles turned seawater near the seafloor extremely murky, and prodigious streams of superhot water poured from the volcanic rocks, where lava flows scorched tubeworms that had not yet decayed. A few partially covered animal colonies still clung to life, while hordes of crabs fed on the carcasses of dead animals.

A huge undersea eruption on the Juan de Fuca Ridge about 250 miles off the Oregon coast poured out batches of lava, creating new oceanic crust in a single convulsion. The ridge forms a border between the huge Pacific plate to the west and the tiny Juan de Fuca plate to the east (Fig. 10–8). Eruptions along the ridge occur when the two plates separate, allowing molten rock from the mantle to rise to the seafloor surface, forming new crust. Over time, the process of seafloor spreading carries older oceanic crust away from the ridge.

The young volcanic rocks included pillow lavas and shiny, bare basalt lacking any sediment cover. Water warmed to 50 degrees Celsius seeped out of cracks in the freshly hardened basalt. In some places, tubeworms had already established residency around thermal vents. The eruption appeared to be related to two megaplumes discovered in the late 1980s. A string of new basaltic mounds more than 10 miles long erupted on a fracture running between the sites of the two megaplumes. The hot hydrothermal fluids, along with fresh basalt, gush out of the ocean floor when the ridge system cracks open and churns out more new crust.

A field of seafloor geysers off the coast of Washington State expels into the near-freezing ocean hot brine at temperatures approaching 400 degrees Celsius. Massive undersea volcanic eruptions from fissures on the ocean floor at spreading centers along the East Pacific Rise create large megaplumes of hot water. The megaplumes result from short bursts of intense volcanic activity and can measure up to several tens of miles wide.

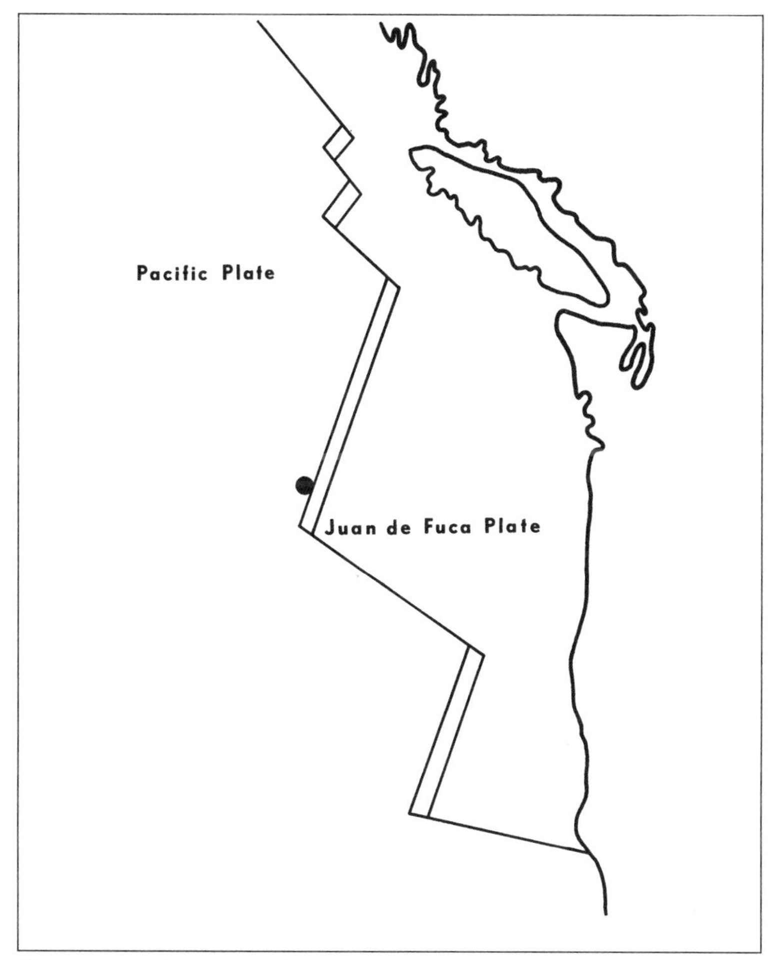

Figure 10–8 Location of volcanic site on the Juan de Fuca Ridge.

The ridge apparently splits open and spills out hot water while lava erupts in an act of catastrophic seafloor spreading. In a matter of a few hours, or at most a few days, more than 100 million cubic yards of superheated water gushes from a large fracture in the oceanic crust up to several miles long. When the seafloor ruptures, vast quantities of hot water held under great pressure beneath the surface violently rush out, creating colossal plumes. The release of massive amounts of superheated water beneath the sea might explain why the ocean remains salty.

SUBMARINE SLIDES

The deep sea is not nearly as quiet as it seems. The constant tumbling of seafloor sediments down steep banks churns the ocean bottom into a murky mire. The largest slides occur on the ocean floor, and as many as 40 giant submarine slides are located near U.S. coasts. Submarine slides moving down steep continental slopes have buried undersea telephone cables under a thick layer of rubble. Sediments eroding out from beneath the cable leave it dangling between uneroded areas of the seabed, causing the cable to fail. A modern slide that broke a submarine cable near Grand Banks, south of Newfoundland, moved downslope at a speed of about 50 miles per hour—comparable to large terrestrial slides that can devastate the landscape.

Undersea flow failures also generate large tsunamis that overrun parts of the coast. In 1929, an earthquake on the coast of Newfoundland set off a large undersea slide that triggered a tsunami, killing 27 people. On July 3, 1992, apparently a large submarine slide sent a 25-mile-long, 18-foot-high wave crashing down on Daytona Beach, Florida, overturning automobiles and injuring 75 people.

Submarine slides carve out deep canyons in continental slopes. They consist of dense, sediment-laden water that moves sediments swiftly along the ocean floor. These muddy waters, called turbidity currents (Fig. 10–9), travel down continental slopes and transport immensely large blocks. Turbidity currents are also initiated by river discharge, coastal storms, or other currents. They deposit huge amounts of sediment that build up the continental slopes and the flat-lying abyss below.

Figure 10–9 An underwater river of sediment-laden water, called a turbidity current. Photo by R. F. Dill, courtesy of U.S. Navy

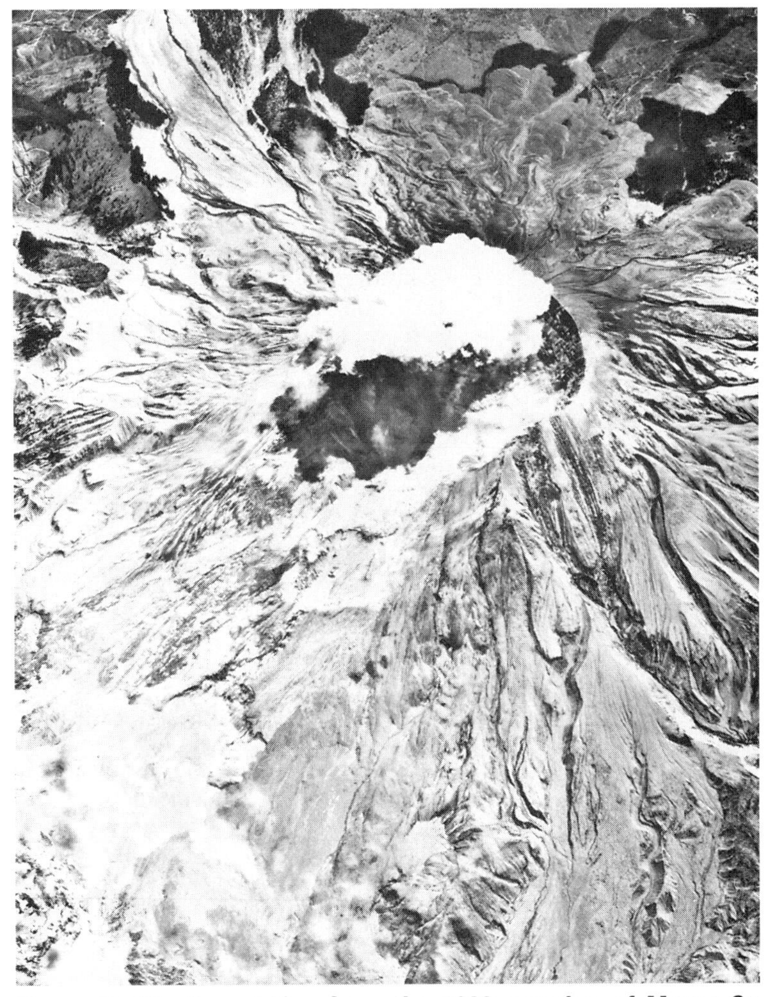

Figure 10–10 Devastation from the 1980 eruption of Mount St. Helens, showing extensive ice and rock debris in the foreground.
Courtesy of NASA

The continental slopes plunge thousands of feet to the ocean floor, inclined at steep angles of 60 to 70 degrees. Sediments reaching the edge of the continental shelf slide off the continental slope by the pull of gravity. Huge masses of sediment cascade down the continental slope by gravity slides, gouging out steep submarine canyons and depositing great heaps of sediment. Such undersea slides are often as catastrophic as landslides and move massive quantities of sediment downslope in a matter of hours.

Submerged deposits near the base of the main island of Hawaii rank among the greatest slides on Earth. On the southeast coast of Hawaii, on Kilauea Volcano's south flank, about 1,200 cubic miles of rock is slumping toward the sea at a breakneck speed in geologic terms—up to 10 inches per year. It is the biggest mass on Earth that is moving in this fashion. Ultimately, catastrophic collapse will occur, far more destructive than any of the volcano's eruptions.

On the ocean floor lies evidence that great chunks of the Hawaiian Islands had once slid into the sea. By far the largest example of an undersea rock slide is along the flank of a Hawaiian volcano. The slide measured roughly 1,000 cubic miles in volume and spread some 125 miles from its point of origin. The collapse of the island of Oahu sent debris 150 miles across the deep-ocean floor, churning the sea into gargantuan waves. When part of Mauna Loa Volcano collapsed and fell into the sea about 100,000 years ago, it created a tsunami 1,200 feet high that was not only catastrophic to Hawaii but might even have caused damage along the Pacific coast of North America.

The bottom of the rift valley of the Mid-Atlantic Ridge holds the remnants of a vast undersea slide at a depth of 10,000 feet, surpassing in magnitude any landslide in recorded history. A large scar on one side of the submarine volcanic range indicates that the mountainside gave way and slid downhill at a tremendous speed, running up and over a smaller ridge farther downslope in a matter of minutes. The slide carried nearly 5 cubic miles of rock debris, or six times more than the 1980 Mount St. Helens landslide, the largest in modern history (Fig. 10–10). The slide appears to have occurred about 450,000 years ago, possibly creating a gigantic sea wave 2,000 feet high.

SEA CAVES

Caves are pounded into existence by ocean waves, plowed open by flowing ice, or arise out of lava flows. They are the most spectacular examples of the dissolving power of groundwater. Over time, acidic water flowing underground dissolves large quantities of limestone, forming a system of large chambers and tunnels. Caves develop from underground channels that carry out water that seeps in from the water table. This creates an underground stream similar in structure to streams on the surface that flow from a breached water table. The limestone landforms resulting from this process are known as karst terrains, after a region in Slovenia famous for its caves.

Caves also develop in sea cliffs (Fig. 10–11) by the ceaseless pounding of the surf or by groundwater flow through an undersea limestone formation hollowed out as the water empties into the ocean. Wave action on limestone promontories with zones of differential hardness creates sea arches. A major storm at sea erodes the tall cliffs landward several tens of feet. Sometimes the pounding of the surf punches a hole in the chalk to form a sea arch.

Figure 10–11 Sea cave cut into siltstone, Chinitna district, Cook Inlet region, Alaska. Photo by A. Grantz, courtesy of USGS

In the jungle on Mexico's Yucatan Peninsula is a bizarre realm of giant caverns linked by miles of twisting passages a hundred feet below the sea. The karst terrain gives birth to underwater caves and sinkholes. The sinkholes form when the upper surface of a limestone formation collapses, exposing the watery world below. The underlying limestone is honeycombed with long tunnels, some several miles in length, and huge caverns that could easily hold several houses.

Just like surface caves, the Yucatan caverns contain a profusion of icicle-shaped formations of stalactites hanging from the ceiling and stalagmites rising from the floor. The formations also include delicate, hollow stalactites called soda straws that took millions of years to create. Fish, crustaceans, and other small, primitive creatures, blind as a result of generations without light, making eyes useless, live in the darkest recesses of the caves.

Blue holes are submerged sinkholes in the sea that appear dark blue because of their great depth. Many blue holes dot the shallow waters surrounding the Bahama Islands southwest of Florida. They formed during the last ice age, when the ocean fell by several hundred feet, exposing parts of the ocean floor well above sea level. The sea lowered in response to the growing ice sheets that covered the northern regions of the world, locking up huge quantities of the world's water.

Figure 10–12 Possibly the nation's largest sinkhole, measuring 425 feet long, 350 feet wide, and 150 feet deep, in Shelby County, central Alabama. Courtesy of USGS

During its exposure on dry land, acidic rainwater seeping into the seabed dissolved the limestone bedrock, creating immense subterranean caverns. Under the weight of the overlying rocks, the roofs of the caverns collapsed, forming huge gaping pits (Fig. 10–12). At the end of the ice age, when the ice sheets melted, the sea inundated the area and submerged the sinkholes. Blue holes can be very treacherous because they often have strong eddy currents or whirlpools that are particularly hazardous to small boats.

SEAFLOOR CRATERS

Because water covers over 70 percent of the Earth's surface, most meteorites land in the ocean, and several sites on the seafloor are possible marine impact craters. An asteroid or comet landing in the ocean would produce a conical-shaped curtain of water as billions of tons of seawater splashes high into the air. The atmosphere would become oversaturated with water vapor, and thick cloud banks would shroud the planet, blocking out the sun. Massive tsunamis would race outward from the impact site and traverse clear around the world. When striking seashores, they would travel hundreds of miles inland, devastating everything in their paths. About 65 million years ago, a large meteorite supposedly struck the Earth, creating a crater at least 100 miles wide whose debris sent the planet into environmental chaos. This catastrophe might have caused the demise of the dinosaurs along with 70 percent of all other species. The actual crater has yet to be found, suggesting that the meteorite landed in the ocean. If so, millions of years of seafloor subduction would have erased all signs of it.

Much of the search for the dinosaur-killer impact site has been concentrated around the Caribbean area (Fig. 10–13), where thick accumulations of wave-deposited rubble exist along with melted and crushed rock apparently ejected from a crater. The most suitable site for the crater is the 110-mile-wide Chicxulub structure, the largest known on Earth. It lies

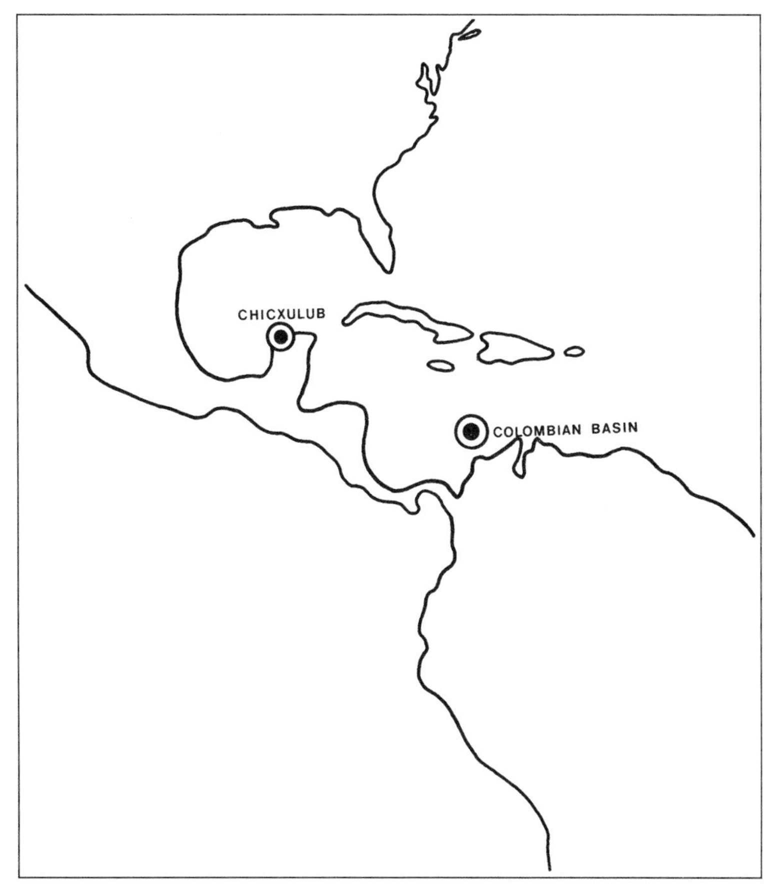

Figure 10–13 Locations of possible impact structures in the Caribbean area that might have ended the Cretaceous period.

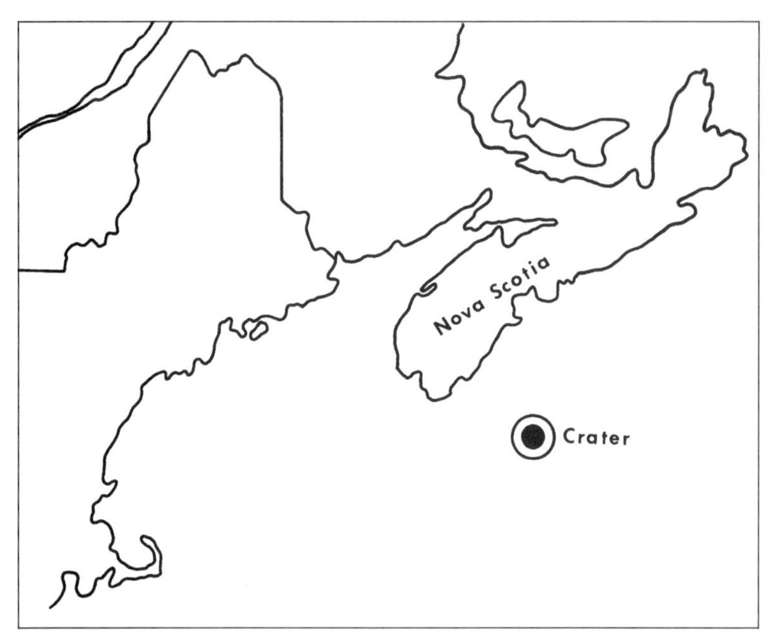

Figure 10–14 Location of the Montagnais crater off Nova Scotia, Canada.

beneath 600 feet of sedimentary rock on the northern coast of the Yucatan Peninsula. If the meteorite landed on the seabed just offshore, 65 million years of sedimentation would have long since buried it under thick deposits of sand and mud. Furthermore, a splashdown in the ocean would have created an enormous sea wave, or tsunami, that would have scoured the seafloor and deposited its rubble on nearby shores.

The most pronounced undersea impact crater known is the 35-mile-wide Montagnais structure 125 miles off the southeast coast of Nova Scotia (Fig. 10–14). Oil companies exploring for petroleum in the area discovered the circular formation. The crater is 50 million years old and closely resembles craters on dry land, except that its rim is 375 feet beneath the sea and the crater bottom is 9,000 feet deep. A large meteorite up to 2 miles wide excavated the crater. The impact raised a central peak similar to those seen inside craters on the moon.

The impact structure also contained rocks melted by a sudden shock. Such an impact would have sent a tremendous sea wave crashing down on nearby shores. Because of its size and location, the crater was thought to be a likely candidate for the source of tektites (small, glassy bodies) strewn across the American West. Unfortunately, its age is several million years too young to have created the North American tektites. However, the ocean is vast, and better candidates might some day reveal themselves.

A meteorite slamming into the Atlantic Ocean along the Virginia coast about 40 million years ago released a huge wave that pounded the adjacent shoreline. Apparently, the tsunami gouged out of the seafloor a 5,000-square-mile region about the size of Connecticut. When the meteorite crashed into the submerged continental shelf, it created a wave that ripped the seafloor into an enormous number of large boulders. A 200-foot-thick layer of 3-foot boulders was deposited in three locations, buried under 1,200 feet of sediment. Within the boulder layer are mineral grains showing shock features and glassy rocks called tektites that formed when a meteorite blasted the seafloor and flung the molten rock in all directions.

A large meteorite impact might have created the Everglades at the southern tip of Florida. The Everglades is a swamp and forested area surrounded by an oval-shaped system of ridges upon which rests most of southern Florida's cities. A giant coral reef, dating about 6 million years old, lies beneath the rim surrounding the Everglades. The coral reef probably formed around the circular basin gouged out by the meteorite impact. A thick layer of limestone surrounding the area and laid down about 40 million years ago is suspiciously missing over most of the southern part of the Everglades. Apparently, a large meteorite slammed into limestones submersed under 600 feet of water and fractured the rocks. The impact also would have generated an enormous tsunami and swept the debris far out to sea.

About 2.3 million years ago, a major asteroid appears to have impacted on the ocean floor in the Pacific Ocean roughly 700 miles west of the tip of South America. Although no crater has been found, an excess of iridium (a rare isotope of platinum found in abundance on meteorites) in sand-size bits of glassy rock existed in the area, suggesting an extraterrestrial origin. The impact created at least 300 million tons of debris, consistent with an object about a half mile in diameter. The blast from the impact would have been equal to that of all the nuclear arsenals in the world today, with devastating consequences for the local ecology. Moreover, geologic evidence suggests that the Earth's climate cooled dramatically between 2.2 and 2.5 million years ago, when glaciers covered large parts of the Northern Hemisphere.

UNDERSEA EXPLOSIONS

The most explosive volcanic eruption in recorded history occurred during the 17th century B.C. on the island of Thera, 75 miles north of Crete in the Mediterranean Sea. The magma chamber beneath the island apparently flooded with seawater, and like a gigantic pressure cooker the volcano blew its lid. The volcanic island collapsed into the emptied magma chamber, forming a deep water-filled caldera that covered an area of 30 square miles (Fig. 10–15). The collapse of Thera also created an immense sea wave that battered the shores of the eastern Mediterranean.

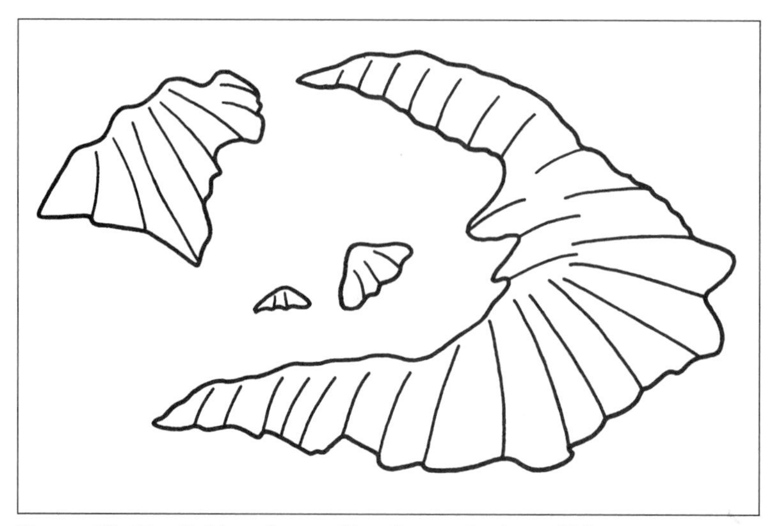

Figure 10–15 Caldera formed by the explosion of Thera.

Krakatoa lies in the Sundra Strait between Java and Sumatra, Indonesia. On August 27, 1883, a series of four powerful explosions ripped the island apart. The explosions were probably powered by the rapid expansion of steam, generated when seawater entered a breach in the magma chamber. Following the last convulsion, most of the island caved into the emptied magma chamber and created a large undersea caldera more than 1,000 feet below sea level, resembling a broken bowl of water with jagged edges protruding above the surface of the sea.

The first hydrogen bomb test was conducted on November 1, 1952, on Elugelab atoll, in the Eniwetok lagoon in the South Pacific. The nuclear device was named "Mike" and measured 22 feet long, 5 feet wide and weighed about 65 tons, with an explosive force estimated at 10 megatons of TNT. When Mike was detonated (Fig. 10–16), the fireball expanded to more than 3 miles in diameter in less than a second. Millions of gallons of seawater instantly boiled into steam. After the clouds cleared, Elugelab was no more. A huge crater was blown in the ocean floor 1 mile wide and 1,500 feet deep.

Another type of crater on the bottom of the ocean forms by a natural seafloor explosion. In 1906, sailors in the Gulf of Mexico witnessed a massive gas blowout that sent mounds of bubbles to the surface. The area is known for its reservoirs of hydrocarbons that might have caused the explosion. Pockets of gases lie trapped under high pressure deep beneath the floor of the ocean. As the pressure increases, the gases explode undersea, spreading debris in all directions and producing huge craters on the ocean floor. The gases rush to the surface in great masses of bubbles that burst in the open air, resulting in a thick foamy froth on the surface of the ocean.

Further exploration of the site yielded a large crater on the ocean floor, lying in 7,000 feet of water southeast of the Mississippi River Delta. The elliptical hole measured 1,300 feet long, 900 feet wide, and 200 feet deep and sat atop a small hill. Downslope lay more than 2 million cubic yards of ejected sediment. Apparently, gases seeped upward along cracks in the seafloor and collected under an impermeable barrier. Eventually, the pressure forced the gas to blow off its cover, forming a huge blowout crater.

In the Gulf of Mexico, as well as in other parts of the world, the seabed overlies thick salt deposits formed when the sea evaporated during a warmer climate. The bottom of the Gulf is lined with a layer of anhydrite, an anhydrous (water-saturated) calcium sulfate common in evaporite de-

Figure 10–16 The test of the first hydrogen bomb, "Mike," on Elugelab Atoll on November 1, 1955. Courtesy of Defense Nuclear Agency

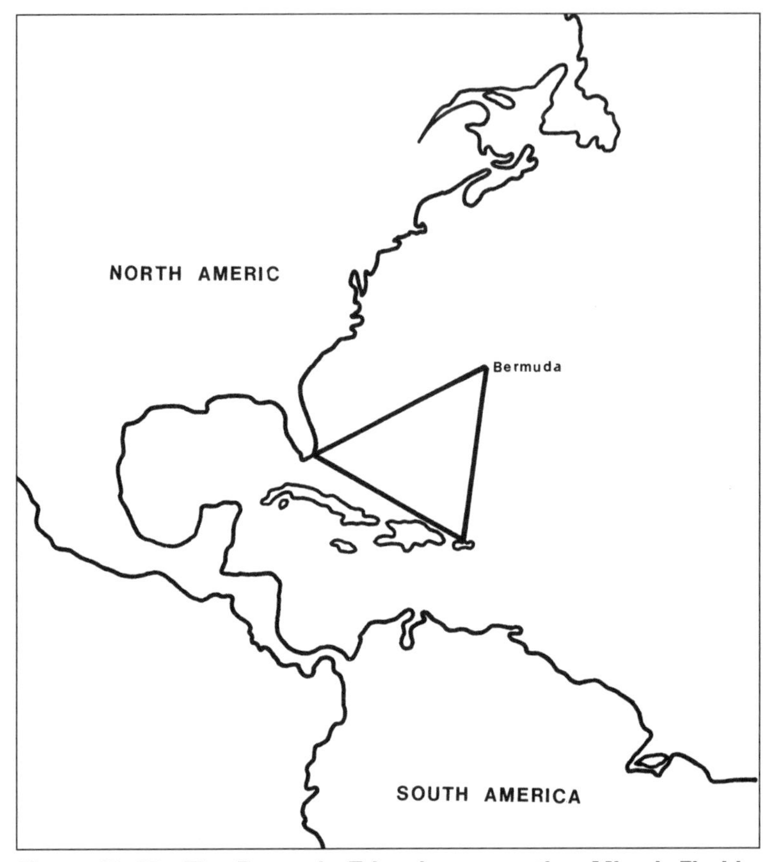

Figure 10–17 The Bermuda Triangle, connecting Miami, Florida, Puerto Rico, and Bermuda, has been blamed for mysterious disappearances of ships and planes.

posits. The anhydrite forms an impervious stratum to the buildup of gas beneath the surface.

When the building gas pressure overcomes the barrier, gases rush toward the surface, forming a froth on the open ocean. A ship sailing into such a foamy sea would suddenly lose all buoyancy because it would no longer be supported by seawater, and would immediately sink to the bottom. An airplane flying overhead might stall out, its engines choking on the pall of poisonous gases. Perhaps these phenomena might explain the strange disappearances of ships and aircraft in the Caribbean around a region known as the Bermuda Triangle (Fig. 10–17), one of many unsolved mysteries of the sea.

GLOSSARY

abrasion	erosion by friction caused by rock particles carried by running water, ice, and wind
absorption	the process by which radiant energy—incident on any substance—is retained and converted into heat or other forms of energy
abyss	the deep ocean, generally over 1 mile in depth
accretion	the increase in the size of an object by the accumulation of material to its exterior
advection currents	the horizontal movement of air or water
albedo	the amount of sunlight reflected from an object, determined by its color and texture
alluvium	stream-deposited sediment
alpine glacier	a mountain glacier or a glacier in a mountain valley
andesite	a volcanic rock generally associated with subduction zone volcanoes
arthropods	a large group of invertebrates, including crustaceans and insects
asthenosphere	a layer of the upper mantle—roughly between 50 and 200 miles below the surface—that is more plastic than

	the rock above and below and might be in convective motion
atmospheric pressure	the weight per unit area of the total mass of air above a given point; also called barometric pressure
backarc basin	a seafloor spreading system of volcanoes caused by plate extension behind an island arc above a subduction zone
Baltica	an early Paleozoic European continent
barrier island	a low, elongated coastal island that parallels the coastline and protects the shore from storms
basalt	a volcanic rock that is dark in color and usually quite fluid in the molten state
Benioff zone	a plane of seismic activity outlining the boundaries of a plate subducting into the mantle
benthic front	ocean currents that flow along the bottom of the deep ocean
biogenic	sediments composed of the remains of plant and animal life, such as shells
biosphere	the living portion of the Earth that interacts with all other biological and geologic processes
black smoker	superheated hydrothermal water rising to the surface at a midocean ridge. The water is supersaturated with metals, and when exiting through the seafloor, the water quickly cools and the dissolved metals precipitate, resulting in black, smokelike effluent
brachiopods	shallow-water, marine invertebrates with bivalve shells similar in appearance to clams
caldera	a large depression at the summits of volcanoes formed by explosive eruptions or collapse
calving	formation of icebergs by the breaking off of glaciers entering the ocean
carbonaceous	a substance containing carbon, such as sedimentary rocks and meteorites

carbonate	a mineral containing calcium carbonate, such as in limestone
carbon cycle	the flow of carbon into the atmosphere and ocean, the conversion to carbonate rock, and the return by volcanoes
Cenozoic era	the age of recent life, spanning from 65 million years ago to the present
chemosynthesis	manufacturing of organic compounds by energy from chemical reactions, such as those on the deep-sea floor near hydrothermal vents
circum-Pacific belt	active seismic regions on the rim of the Pacific plate, coinciding with the Ring of Fire
climate	the average course of the weather for a certain region over time
coastal storm	a cyclonic, low-pressure system moving along a coastal plain or just offshore. It causes north to northeast winds over the land and, along the Atlantic seaboard, it is called a northeaster
condensation	the process whereby a substance changes from the vapor phase to liquid or solid phase; the opposite of evaporation
continent	a landmass composed of light granitic rock riding on the denser rocks of the upper mantle
continental drift	the concept that the continents have been drifting across the surface of the Earth throughout geologic time
continental glacier	an ice sheet covering a portion of a continent
continental margin	the true edge of a continent, usually undersea
continental shelf	the offshore area of a continent in shallow sea
continental slope	the transition from the continental margin to the deep sea basin

convection	a circular, vertical flow of a fluid medium due to heating from below. As materials are heated they become less dense and rise, while cooler, heavier materials sink
convergent plate boundary	the boundary between lithospheric plates where one plate subducts under another in deep-sea trenches
coral	any of a large group of shallow-water, bottom-dwelling marine invertebrates, which commonly build reef colonies in warm waters
core	the inner Earth with a radius of 2,300 miles consisting of a crystalline inner core and molten outer core of iron and nickel. Also, a rock sample taken vertically into the Earth
Coriolis effect	the apparent force that deflects air or ocean currents, causing them to curve in relation to the rotating Earth
craton	the stable interior of a continent, usually composed of the oldest rocks on the continent
crust	the outer layers of the Earth's rocks
delta	the wedge-shape pile of sediments deposited at the mouth of a river
density	the amount of any quantity per unit volume
dew point	the temperature to which air at a constant pressure and moisture content must be cooled for saturation to occur
diapir	the buoyant rise of a rock toward the surface through heavier rock
diatom	microplants, whose silica fossil shells form siliceous sediments called diatomite
divergent plate boundary	the boundary between lithospheric plates where they separate at midocean ridges
downwelling	the sinking of a fluid because it is heavier than the surrounding medium
earthquake	the sudden rupture of rocks along active faults
East Pacific Rise	a midocean spreading center that runs north-south along the eastern side of the Pacific. The predominant location

upon which hot springs and black smokers have been discovered

echinoderms — marine invertebrates, including starfish, sea urchins, and sea cucumbers

El Niño — an anomalous warning in the eastern equatorial Pacific due to the breakdown of the trade winds

evaporation — the transformation of a liquid to a gas

evaporite — the deposition of salt, anhydrite, and gypsum from evaporation in an enclosed basin of stranded seawater

evolution — the tendency of physical and biologic factors to change with time

extrusion — the ejection of lava onto the Earth's surface from below

fissure — a large crack in the crust through which magma escapes to the surface

fluvial — pertaining to being deposited by a river

foraminifera — calcium carbonate secreting organisms that live in the surface waters of the oceans. After death, their shells form the primary constituents of limestone and sediments deposited on the seafloor

fossil — any remains, impression, or trace in rock of a plant or animal of a previous geologic age

fossil fuel — an energy source derived from ancient plant and animal life, including coal and petroleum

fracture zones — narrow regions consisting of ridges and valleys parallel to spreading ridges and aligned in a stairstep shape

frequency — the rate at which crests of any wave pass a given point

gabbro — a coarse-grained, intrusive rock common in oceanic crust, which is rich in iron and magnesium

gastropod — a large class of mollusks, including slugs and snails

geologic column — the total thickness of rock units in a region

geostrophic flow	ocean currents that flow perpendicular to the Coriolis flow or to the right of the boundary currents in the Northern Hemisphere
geyser	a hot spring that periodically flashes to steam and blows out at the surface
gneiss	a metamorphic rock derived from granite subjected to high temperatures and pressures
Gondwana	a southern supercontinent of Paleozoic time, consisting of Africa, South America, India, Australia, and Antartica. It broke up into present continents during the Mesozoic era
granite	a coarse-grained, silica-rich rock consisting primarily of quartz and feldspars. It is the principal constituent of the continents and is believed to be derived from a molten state beneath the Earth's surface
gravimeter	an instrument that measures the intensity of the Earth's gravity
greenhouse effect	the global heating effect due to the atmosphere being more transparent to incoming short-wave solar radiation than to outgoing long-wave radiation
greenstone belt	a mass of Precambrian, metamorphosed igneous rock
guyot	an undersea volcano that reached the surface of the ocean, whereupon its top was flattened by erosion. Later, subsidence caused the volcano to sink below the surface preserving its flat top appearance
heat flow	heat energy flows from hot areas toward cold at a rate or flux equal to the temperature gradient times the conductivity of the material in between
hot spot	a volcanic center with no relation to plate boundary location; an anomalous magma generation site in the mantle
hydrocarbon	a molecule consisting of carbon chains with attached hydrogen atoms
hydrogenous	pertaining to sedimentary deposits from the water column

hydrosphere	the water layer at the surface of the Earth
hydrothermal	the convection of cold seawater downward and circulation through the oceanic crust toward the deeper depths where it becomes hot and is buoyantly forced back toward the surface
Iapetus Sea	a former sea that occupied an area similar to the present Atlantic Ocean, prior to the assemblage of Pangaea
ice age	a period of time when large areas of the Earth were covered with glaciers
iceberg	a portion of a glacier calved off a landmass into the sea
ice cap	a polar cover of ice and snow
igneous rocks	all rocks that have solidified from a molten state
impact	a point on the surface of a planetary body where a celestial object lands to create a crater
internal wave	a wave propagating at a density boundary within the ocean rather than at the surface of the water
interstitial	relating to intergranular spaces in sediments
invertebrates	animals with external skeletons, such as shellfish
iridium	a rare isotope of platinum, relatively abundant on meteorites
island arc	the volcanoes landward of a subduction zone, parallel to the submarine trench, and above the melting zone of a subduction zone
landslide	rapid downhill movement of earth materials often triggered by earthquakes and storms
Langmuir circulation	near-surface alternating vortices aligned downwind, generated by the interaction of waves and mean shear
Laurasia	the northern supercontinent of the Paleozoic, consisting of North American, Europe, and Asia
Laurentia	a Precambrian continent comprising North America, Greenland, and northern Europe
lava	molten magma that flows out onto the surface

limestone	a sedimentary rock composed of calcium carbonate that is secreted from seawater by invertebrates, whose skeletons make up the bulk of the deposits
lithosphere	a rigid outer layer of the mantle, typically about 60 miles thick. It is overridden by the continental and oceanic crusts and is divided into segments called plates
lysocline	the ocean depth below which the rate of dissolution exceeds the rate of deposition of the dead shells of calcareous organisms
magma	a molten rock material generated within the Earth that is the constituent of igneous rocks, including volcanic eruptions
magnetic field reversal	a reversal in the polarity of the Earth's magnetic poles. This has occurred intermittently throughout geologic time
magnetometer	a devise used to measure the intensity and direction of the magnetic field
manganese nodule	a cobble-shaped ore on the deep-sea floor, which is rich in manganese and iron
mantle	the part of the Earth below the crust and above the core, composed of dense, iron and magnesium-rich rocks
massive sulfide	an ore deposit in which sulfur chemically combines with metals
megaplume	a large volume of mineral-rich warm water above an oceanic rift
Mesozoic	the age of middle life; refers to the period between 250 and 65 million years ago
metamorphic rock	a rock crystallized from previous igneous, metamorphic, or sedimentary rocks created under conditions of intense temperatures and pressures without melting
meteorite	a metallic or stony body from space that enters the atmosphere and impacts on the Earth's surface
microplate	a small block of oceanic crust surrounded by major plates

Mid-Atlantic Ridge	the seafloor spreading ridge of volcanoes that marks the extensional edge if the North American and South American plates to the west and the Eurasian and African plates to the east
midocean ridge	a submarine ridge along a divergent plate boundary where new ocean floor is created by the upwelling of mantle material
moho	the boundary between the crust and mantle
mollusks	a large group of invertebrates, including clams, snails, squids, and extinct ammonites
olivine	a magnesium iron silicate mineral that is the primary constituent of the mantle
ophiolite	a mass of oceanic crust thrust onto the continents by plate collisions
ore body	the accumulation of metal-bearing ores, where the hot hydrothermal water moving upward toward the surface mixes with cold seawater penetrating downward
orogeny	an episode of mountain building by tectonic activity
outgassing	the loss of gas within a planet, as opposed to degassing or loss of gas from meteorites
oxidation	the chemical combination of oxygen with other elements
paleomagnetism	the study of the Earth's magnetic field, including the position and polarity of the poles in the past
paleontology	the study of ancient life forms based on the fossil record of plants and animals
Paleozoic	the age of ancient life, between 570 and 250 million years ago
Pangaea	an ancient supercontinent that included all the landmasses of the Earth
Panthalassa Sea	the great world ocean that surrounded Pangaea
peridotite	the most common ultramafic rock type in the Earth's mantle

pH scale	a logarithmic scale depicting the acidity or alkalinity of a substance. A pH of 0 is the strongest acid; a pH of 14 is the strongest base; and a pH of 7 is neutral
photosynthesis	the process by which plants create carbohydrates from carbon dioxide, water, and sunlight
phytoplankton	marine plant microorganisms
pillow lava	lava extruded on the ocean floor giving rise to tubular shapes
plate tectonics	the theory that accounts for the major features of the Earth's surface in terms of the interaction of lithospheric plates
precipitation	products of condensation that fall from clouds as rain, snow, hail, or drizzle; also the deposition of minerals from seawater
primary producer	the lowest member of a food chain
primordial	referring to the first created or developed
radiogenic	pertaining to heat generated by radioactive sources
radiolarians	siliceous-shelled microorganisms whose shells make up a large component of siliceous sediments
radiometric dating	the determination of how long an object has existed by chemical analysis of stable versus unstable radioactive elements
reef	the biological community that lives at the edge of an island or continent. The shells form limestone preserved in the geologic record
regression	a fall in sea level, exposing continental shelves to erosion
resource	useful earth materials such as metallic ores
ridge crest	an axis of midocean volcanoes aligned along the edge of two plates extending away from each other
rift valley	the center of an extensional spreading center where continental or oceanic plate separation occurs

Ring of Fire	a belt of subduction zones around the Pacific plate related to volcanic activity
sandstone	a sedimentary rock consisting of cemented sand grains
seafloor spreading	a process by which the ocean floor is created by the separation of lithospheric plates along the midocean ridges, with new oceanic crust formed from mantle material which rises from the mantle to fill the rift
seamount	an undersea volcano that never reached the surface of the ocean and so does not have a flat erosional top
seawall	a structure built to protect against shore erosion
seaward bulge	the elevated bulge produced by the bending of a subducting plate
sedimentation	the deposition of sediments
seiche	the oscillation of water in a bay
seismic	relating to earthquake or artificial vibrations
seismic sea wave	an ocean wave related to an undersea earthquake, also called a tsunami
serpentine	a low-grade metamorphic rock derived from hydrated olivine
shield	areas of the exposed Precambrian nucleus of a continent
shield volcano	a broad, low-lying volcanic cone built up by lava flows of low viscosity
sonar	an instrument for measuring the ocean floor with sound waves
sounding	the measurement of water with a weighted line
storm surge	an abnormal rise of the water level along a shore as a result of a coastal storm
subduction zone	an area where an oceanic plate dives below a continental plate into the asthenosphere. Ocean trenches are the surface expressions of subduction zones

submarine canyon	a deep gorge in the seabed formed by the undersea extensions of rivers
subsidence	the collapse of sediments due to the removal of undergound fluids
surge glacier	a continental glacier heading toward the sea at a high rate of advance during certain times
symbiosis	the union of two dissimilar organisms for mutual benefit
tectonic activity	the formation of the Earth's crust by large-scale earth movements throughout geologic time
tephra	all clastic material—from dust particles to large blocks—expelled from volcanoes during eruptions
terrane	a section of oceanic crust that collided with a continent and is of different composition from surrounding rocks
Tethys Sea	the hypothetical mid-latitude area of the oceans separating the northern and southern continents of Gondwana and Laurasia
thermocline	the boundary between cold and warm layers of the ocean
tidal friction	the loss of energy through heating caused by the movements associated with the tides
tide	a bulge in the ocean produced mainly by the moon's gravitational forces on the Earth's oceans. The rotation of the Earth beneath this bulge causes the rising and lowering of sea level
transform fault	a major plate boundary formed when two plates move across each other along a fault
transgression	a rise in sea level that causes flooding of the shallow edges of continental margins
traps	a succession of lava flows in a stairstep shape
trench	a topographic feature formed when the seafloor plunges into the mantle along a line of subduction
tsunami	a sea wave generated by an undersea earthquake or volcanic eruption

tubeworm	a retractable wormlike animal living within a long stalk near hydrothermal vents
turbidite	a slurry of mud that periodically slides down often gentle slopes toward the deep-sea floor
typhoon	a severe tropical storm in the Western Pacific similar to a hurricane
underplating	adhesion of mantle material to the underside of a plate causing it to thicken
upwelling	the upward convection of water currents
volcanism	any type of volcanic activity
volcano	a fissure or vent in the crust through which molten rock rises to the surface to form a mountain
white smoker	a hydrothermal vent on the deep-sea floor similar to a black smoker but which produces a white effluent

BIBLIOGRAPHY

The Blue Planet

Allegre, Claud J. and Stephen H. Snider. "The Evolution of the Earth." *Scientific American* 271 (October 1994): 66–75.

Barnes-Svarney, Patricia. "In Search of Ancient Shores." *Earth Science* 40 (Spring 1987): 22.

Gould, Stephen Jay. "The Evolution of Life on the Earth." *Scientific American* 271 (October 1994): 85–91.

Horgan, John. "In the Beginning." *Scientific American* 264 (February 1991): 117–125.

Kasting, James F., Owen B. Toon, and James B. Pollack. "How Climates Evolved on the Terrestrial Planets." *Scientific American* 258 (February 1988): 90–97.

Kerr, Richard A. "An About-Face Found in the Ancient Ocean." *Science* 253 (September 20, 1991): 1359–1360.

Knauth, Paul. "Ancient Sea Water." *Nature* 362 (March 25, 1993): 290–291.

Nance, R. Damian, Thomas R. Worsley, and Judith B. Moody. "The Supercontinent Cycle." *Scientific American* 259 (July 1988): 72–79.

Taylor, Stuart Ross. "Young Earth like Venus." *Nature* 350 (April 4, 1991): 376–377.

Vermeij, Geerat J. "The Biological History of a Seaway." *Science* 260 (June 11, 1993): 1603–1604.

Waldrop, M. Mitchell. "Goodbye to the Warm Little Pond?" *Science* 250 (November 23, 1990): 1078–1080.

Maritime Exploration

Bartusiak, Marcia. "Mapping the Sea Floor from Space." *Popular Science* 224 (February 1984): 81–85.

Broad, William J. "Life Springs Up in Ocean's Volcanic Vents, Deep Divers Find." *The New York Times* (October 19, 1993): C4.

Broecker, Wallace S. "The Ocean." *Scientific American* 249 (September 1983): 146–160.

Hoffman, Kenneth A. "Ancient Magnetic Reversals: Clues to the Geodynamo." *Scientific American* 258 (May 1988): 76–83.

Kerr, Richard A. "Coming up Short in Crustal Quest." *Science* 254 (December 6, 1991): 1456–1457.

———. "Shaping New Tools for Paleoceanographers." *Science* 234 (October 20, 1986): 427–428.

Monastersky, Richard. "Rare Rocks Drilled from Pangaean Time." *Science News* 134 (August 13, 1988): 166.

Weisburd, Stefi. "Sea-Surface Shape by Satellite." *Science News* 129 (January 18, 1986): 37.

Wood, Dennis. "The Power of Maps." *Scientific American 268* (May 1993): 89–93.

Zimmer, Carl. "Inconsistent Field." *Discover* 15 (February 1994): 26–27.

The Dynamic Seafloor

Berner, Robert A. and Antonio C. Lasaga. "Modeling the Geochemical Carbon Cycle." *Scientific American* 260 (March 1989): 74–81.

Cann, Joe and Cherry Walker. "Breaking New Ground on the Ocean Floor." *New Scientist* 139 (October 30, 1993): 24–29.

Cathles, Lawrence M., III. "Scales and Effects of Fluid Flow in the Upper Crust." *Science* 248 (April 20, 1990): 323–328.

Francheteau, Jean. "The Oceanic Crust." *Scientific American* 249 (September 1983): 114–129.

Gordon, Richard G. and Seth Stein. "Global Tectonics and Space Geodesy." *Science* 256 (April 17, 1992): 333–341.

Howell, David G. "Terranes." *Scientific American* 252 (November 1985): 116–125.

Kerr, Richard A. "Puzzling Out the Tectonic Plates." *Science* 247 (February 16, 1990): 808.

———. "Ocean Crust Role in Making Seawater." *Science* 239 (January 15, 1988): 260.

Monastersky, Richard. "Drilling Shortcut Penetrates Earth's Mantle." *Science News* 143 (February 20, 1993): 117.

Mutter, John C. "Seismic Images of Plate Boundaries." *Scientific American* 254 (February 1986): 66–75.

Ridges and Trenches

Bonatti, Enrico and Kathlene Crane. "Oceanic Fracture Zones." *Scientific American* 250 (May 1984): 40–51.

Green, Harry W., II. "Solving the Paradox of Deep Earthquakes." *Scientific American* 271 (September 1994): 64–71.

Gurnis, Michael. "Ridge Spreading, Subduction, and Sea Level Fluctuations." *Science* 150 (November 16, 1990): 970–972.

Kerr, Richard A. "Having it Both Ways in the Mantle." *Science* 258 (December 1992): 1576–1578.

Macdonald, Kenneth C. and Paul J. Fox. "The Mid-Ocean Ridge." *Scientific American* 262 (June 1990): 72–79.

Monastersky, Richard. "Mid-Atlantic Ridge Survey Hits Bull's-Eye." *Science News* 135 (May 13, 1989): 295.

———. "The Whole-Earth Syndrome." *Science News* 133 (June 11, 1988): 378–380.

Peacock, Simon M. "Fluid Processes in Subduction Zones." *Science* 248 (April 20, 1990): 329–336.

Powell, Corey S. "Peering Inward." *Scientific American* 264 (June 1991): 100–111.

Sullivan, Walter. "Earth's Crust Sinks Deep, Only to Rise in Plumes of Lava." *The New York Times* (June 15, 1993): C1 & C8.

Wickelgren, Ingrid. "Simmering Planet." *Discover* 11 (July 1990): 73–75.

Zimmer, Carl. "The Ocean Within." *Discover* 15 (October 1994): 20–21.

Submarine Volcanoes

Berreby, David. "Barry Versus the Volcano." *Discover* 12 (June 1991): 61–67.

Coffin, Millard F. and Olav Eldholm. "Large Igneous Provinces." *Scientific American* 269 (October 1993): 42–49.

Courtillot, Vincent E. "A Volcanic Eruption." *Scientific American* 263 (October 1990): 85–92.

Hekinian, Roger. "Undersea Volcanoes." *Scientific American* 251 (July 1984): 46–55.

Kerr, Richard A. "Did Pinatubo Send Climate-Warming Gases into a Dither?" *Science* 263 (March 18, 1994): 1562.

Monastersky, Richard. "Garden of Volcanoes in the Pacific." *Science News* 143 (June 5, 1993): 367.

———. "Set Adrift by Wandering Hotspots." *Science News* 132 (October 17, 1987): 250–252.

Rampino, Michael R. and Richard B. Stothers. "Flood Basalt Volcanism During the Past 250 Million Years." *Science* 241 (August 5, 1988): 663–667.

Richardson, Randall M. "Bermuda Stretches a Point." *Nature* 350 (April 25, 1991): 655.

Vink, Gregory E., W. Jason Morgan, and Peter R. Vogt. "The Earth's Hot Spots." *Scientific American* 252 (April 1985): 50–57.

White, Robert S. and Dan P. McKenzie. "Volcanism at Rifts." *Scientific American* 261 (July 1989): 62–71.

Abyssal Currents

Folger, Tim. "Waves of Destruction." *Discover* 15 (May 1994): 68–73.

Garrett, Chris. "A Stirring Tale of Mixing." *Nature* 364 (August 19, 1993): 670–671.

Gordon, Arnold L. and Josefino C. Comiso. "Polynyas in the Southern Ocean." *Scientific American* 258 (June 1988): 90–97.

Hollister, Charles D., Arthur R. M. Nowell, and Peter A. Jumars. "The Dynamic Abyss." *Scientific American* 250 (March 1984): 42–53.

Kerr, Richard A. "Ocean-in-a-Machine Starts Looking Like the Real Thing." *Science* 260 (April 2, 1993): 32–33.

———. "How to Stir Up a Deep-Sea Storm." *Science* 231 (January 10, 1986): 117.

Kunzig, Robert. "The Iron Man's Revenge." *Discover* 15 (June 1994): 32–35.

Lockridge, Patricia A. "Volcanoes and Tsunamis." *Earth Science* 42 (Spring 1989): 24–25.

Monastersky, Richard. "Getting the Drift of Ocean Circulation." *Science News* 144 (August 21, 1993): 117.

Ramage, Colin S. "El Niño." *Scientific American* 254 (June 1986): 77–83.

Thorpe, S. A. "Small-Scale Processes in the Upper Ocean Boundary Layer." *Nature* 318 (December 12, 1985): 519–522.

Coastal Geology

Friedman, Gerald M. "Slides and Slumps." *Earth Science* 41 (Fall 1988): 21–23.

Horgan, John. "Antarctic Meltdown." *Scientific American* 268 (March 1993): 19–28.

Maslin, Mark. "Waiting for the Polar Meltdown." *New Scientist* 139 (September 4, 1993): 36–41.

Meyer, Alfred. "Between Venice and the Deep Blue Sea." *Science 86* 7 (July/August 1986): 50–57.

Monastersky, Richard. "Seawall's Seal of Approval." *Science News* 134 (December 17, 1988): 398.

Noris, Robert M. "Sea Cliff Erosion: A Major Dilemma." *Geotimes* 35 (November 1990): 16–17.

Peltier, W. R. "Global Sea Level and Earth Rotation." *Science* 240 (May 13, 1988): 895–900.

Schaefer, Stephen J. and Stanley N. Williams, "Landslide Hazards." *Geotimes* 36 (May 1991): 20–22.

Zimmer, Carl. "Landslide Victory." *Discover* 12 (February 1991): 66–69.

SEA RICHES

Abelson, Philip H. "Future Supplies of Energy and Minerals." *Science* 231 (February 14, 1986): 657.

Beddington, John R. and Robert M. May. "The Harvesting of Interacting Species in a Natural Ecosystem." *Scientific American* 247 (November 1982): 62–69.

Borgese, Elisabeth Mann. "The Law of the Sea." *Scientific American* 248 (March 1983): 42–49.

Carpenter, Betsy. "Opening the Last Frontier." *U.S. News and World Report* 105 (October 24, 1988): 64–66.

Davis, Ged R. "Energy for Planet Earth." *Scientific American* 263 (September 1990): 55–62.

Penney, Terry R. and Desikan Bharathan. "Power from the Sea." *Scientific American* 256 (January 1987): 86–92.

Rona, Peter A. "Mineral Deposits from Sea-Floor Hot Springs." *Scientific American* 254 (January 1986): 84–92.

Rosenberg, A. A., et al. "Achieving Sustainable Use of Renewable Resources." *Science* 262 (November 5, 1993): 828–829.

Talbot, Christopher J. and Martin P. A. Jackson. "Salt Tectonics." *Scientific American* 257 (August 1987): 70–79.

Woche, Wirtschafts. "Sea Riches: What Future?" *World Press Review* (November 1984): 23–25.

MARINE BIOLOGY

Brown, Barbara E. and John C. Ogden "Coral Bleaching." *Scientific American* 268 (January 1993): 64–70.

Chen, Ingfei. "Great Barrier Reef: A Youngster to the Core." *Science* 138 (December 8, 1990): 367.

Childress, James J., Horst Felbeck, and George N. Somero." Symbiosis in the Deep Sea." *Scientific American* 257 (May 1987): 115–120.

Culotta, Elizabeth. "Is Marine Biodiversity at Risk?" *Science* 263 (February 18, 1994): 918–920.

Eastman, Joseph T. and Arthur L. DeVries. "Antarctic Fishes." *Scientific American* 255 (November 1986): 106–114.

Elmer-Dewitt, Philip. "Are Sharks Becoming Extinct?" *Time* 137 (March 4, 1991): 67.

Goodavage, Maria. "Murky Waters." *Modern Maturity* 32 (August-September 1989): 44–50.

Kerr, Richard A. "From One Coral Many Findings Blossom." Science 248 (June 15, 1990): 1314.

Kunzig, Robert. "Invisible Garden." *Discover* 11 (April 1990): 67–74.

Lewin, Roger. "Life Thrives Under Breaking Ocean Waves." *Science* 235 (March 20, 1987): 1465–1466.

Litter, Mark M. and Diane S. Litter. "Deepest Known Plant Life Discovered on an Uncharted Seamount." *Science* 227 (January 4, 1985): 57–59.

RARE SEAFLOOR FORMATIONS

Alvarez, Walter and Frank Asaro. "An Extraterrestrial Impact." *Scientific American* 263 (October 1990): 78–84.

Broad, William J. "A Voyage Into the Abyss: Gloom, Gold and Godzilla." *The New York Times* (November 2, 1993): C1 & C12.

Edmond, John M. and Karen Von Damm. "Hot Springs on the Ocean Floor." *Scientific American* 248 (April 1983): 78–93.

Fryer, Patricia. "Mud Volcanoes of the Marianas." *Scientific American* 266 (February 1992): 46–52.

Gorman, Christine. "Subterranean Secrets." *Time* 140 (November 30, 1992): 64–67.

Grieve, Richard A. "Impact Cratering on the Earth." *Scientific American* 262 (April 1990): 66–73.

Kerr, Richard A. "Testing an Ancient Impact's Punch." *Science* 263 (March 11, 1994): 1371–1372.

———. "Surging Plate Tectonics Erupts Beneath the Sea." *Science* 250 (December 1990): 21.

Lipske, Mike. "Wonder Holes." *International Wildlife* 20 (January/February 1990): 47–51.

Monastersky, Richard. "Underwater Eruption Detected in Pacific." *Science News* 144 (August 28, 1993): 132–133.

———. "Animals Seared by Deep-Sea Eruptions." *Science News* 140 (December 7, 1991): 372–373.

Weisburd, Stefi. "Asteroid Origin of the Everglades." *Science News* 128 (November 9, 1985): 294–295.

INDEX

Boldface page numbers indicate extensive treatment of a topic. *Italic* page numbers indicate illustrations or captions. Page numbers followed by *m* indicate maps, by *t* indicate tables, and by *g* indicate glossary.

D

E

Y

Z